Student Workbook

"Key Concepts in Math and Mapping"

by Susan Smythe and Mungandi Nasitwitwi

to Accompany

Physical Geography: The Global Environment, Canadian edition

by H.J. de Blij, Peter O. Muller, Richard S. Williams, Jr, Cathy T. Conrad, and Peter Long

OXFORD
UNIVERSITY PRESS

70 Wynford Drive, Don Mills, Ontario M3C 1J9
www.oup.com/ca

Oxford University Press is a department of the University of Oxford.
It furthers the University's objective of excellence in research, scholarship,
and education by publishing worldwide in

Oxford New York
Auckland Cape Town Dar es Salaam Hong Kong Karachi Kuala Lumpur Madrid
Melbourne Mexico City Nairobi New Delhi Shanghai Taipei Toronto

With offices in
Argentina Austria Brazil Chile Czech Republic France Greece Guatemala Hungary
Italy Japan Poland Portugal Singapore South Korea Switzerland
Thailand Turkey Ukraine Vietnam

Oxford is a trade mark of Oxford University Press
in the UK and in certain other countries

Published in Canada by Oxford University Press

Copyright © Oxford University Press Canada 2007

The moral rights of the author have been asserted

Database right Oxford University Press (maker)

First published 2007

All rights reserved. No part of this publication may be reproduced,
stored in a retrieval system, or transmitted, in any form or by any means,
without the prior permission in writing of Oxford University Press,
or as expressly permitted by law, or under terms agreed with the appropriate
reprographics rights organization. Enquiries concerning reproduction
outside the scope of the above should be sent to the Rights Department,
Oxford University Press, at the address above.

You must not circulate this book in any other binding or cover
and you must impose this same condition on any acquirer.

Library and Archives Canada Cataloguing in Publication Data

Smythe, Susan
Student workbook to accompany Physical geography : the global environment, Canadian edition by
H.J. de Blij, Peter O. Muller, Richard S. Williams, Cathy T. Conrad and Peter Long / Susan Smythe
and Mungandi Nasitwitwi.

Supplement to: Physical geography.
ISBN-13: 978-0-19-542520-8   ISBN-10: 0-19-542520-0

1. Physical geography—Problems, exercises, etc. I. Nasitwitwi, Mungandi, 1963– II. Title.

GB55.P49 2005 Suppl. 1   910'.02   C2006-902254-2

1 2 3 4 – 10 09 08 07
This book is printed on permanent (acid-free) paper ∞.
Printed in Canada

Contents

Introduction	1
Working with Numbers: A Supplement to Unit 1	3
An Introduction to Maps: A Supplement to Unit 3	10
Significant Figures and Graphs: A Supplement to Unit 4	21
Earth–Sun Relationships: A Supplement to Unit 5	29
Radiation and the Heat Balance of the Atmosphere: A Supplement to Unit 7	32
Atmospheric and Surface Temperatures: A Supplement to Unit 8	47
Air Pressure and Winds: A Supplement to Unit 9	52
Circulation Patterns of the Atmosphere: A Supplement to Unit 10	56
Moisture in the Atmosphere: A Supplement to Unit 12	60
Minerals and Igneous Rocks: A Supplement to Unit 30	67
Sedimentary and Metamorphic Rocks: A Supplement to Unit 31	78

Working with Topographic Maps

Section A: Locational Systems	81
Section B: Directional Indicators	88
Section C: Contour Lines	90
Section D: Contour Calculations	98

Introduction

This document has two purposes. First, it is meant to supplement *Physical Geography: The Global Environment, Canadian Edition*. It is designed to complement existing material, and in particular, to round out some of the weather and climate units. Second, it is meant to be a student handbook, providing information and coaching to help students acquire hands-on skills and to apply these skills to the conceptual framework they pick up from the text.

Students enter first-year physical geography courses from a wide range of experiences. Those with strong science backgrounds will need little help mastering the quantitative side of the discipline. However, students coming from more a liberal arts background may appreciate some guidance with quantitative skills. This workbook has been written with the latter group in mind.

The first part of the workbook is organized around specific Units from the text. Because students live in a Système Internationale world, but have little working knowledge of it, the workbook begins with an exploration of scientific notation and SI units. Then we break open some of the text material on scale and latitude/longitude to provide a foundation for students taking courses in weather, climate, and/or geomorphology.

The material in the workbook then shifts to focus specifically on weather and climate. It is organized around *Physical Geography*'s units in order to add additional information and quantitative expression of concepts after the topics have been introduced in the text.

The four sections at the end of the workbook support instruction in earth sciences and geomorphology by providing additional material for a student's work with topographical maps. This material does not directly correlate to any specific units within *Physical Geography*, but instead it can be usefully applied to the course as a whole.

Supplement/Handbook Part I

Units in *Physical Geography: The Global Environment*	**Supplement Topics**
Unit 1	- scientific notation - SI units
Unit 3	- transverse Mercator projections - map scale - latitude/longitude and adding/subtracting measures of arc
Unit 4	- significant figures - development and interpretation of graphs
Unit 5	- subsolar point and solar declination - solar altitude calculations

Unit 7		electromagnetic radiation
		the absolute temperature scale
		Wien's Displacement Law
		the Stefan-Boltzmann Law
		total energy emission from a blackbody
		the Inverse Square Law
		the Cosine Law of Illumination
		components of the radiation balance
		reflectivity and transmission
		interpretation of radiation balance data over annual and diurnal timeframes
		components of the energy balance
		interpretation of energy balance data over annual and diurnal timeframes
		atmospheric greenhouse effect
Unit 8		temperature calculations:
		○ mean daily temperature
		○ mean annual temperature range
		○ mean monthly temperature
		○ mean annual temperature
		temperature patterns including:
		○ temperature maps
		○ isotherms
		temperature gradients
Unit 9		elevation and pressure
		pressure gradient
		Ideal Gas Law
Unit 10		scales of atmospheric circulation
		factors inducing pressure systems
Unit 12		measures of moisture in the atmosphere
		adiabatic calculations
Unit 30		minerals
		mineral identification
		igneous rocks
Unit 31		sedimentary rocks
		metamorphic rocks

Supplement/Handbook Part II: Working with Topographical Maps

Section A		locational systems on Canadian topographic maps
Section B		direction indicators on Canadian topographic maps
Section C		working with contour lines on topographic maps
Section D		topographic map calculations

Working with Numbers: A Supplement to Unit 1

First-year physical geography students will work with a variety of data, such as distance, temperature, energy, or area measurements, that vary from extremely small to very large values. The purpose of this unit is to help students who have not taken many science courses build a foundation for conceptualizing and working with quantitative geographic data.

This section contains background information, sample calculations and practice questions on the:

1. use of scientific notation to express small and large numbers, and
2. Système International (SI) system of measurement used in Canada.

Scientific Notation

Expressing small and large numbers in scientific notation

Scientific notation allows us to easily represent and perform calculations with the very large and very small numbers that describe the universe. By convention, numbers expressed in scientific notation follow a particular format: the 'coefficient' is followed by the 'exponent term'. In the coefficient, as illustrated below, one figure is given before the decimal, which is then followed by any remaining figures. The coefficient is multiplied by the exponent term—a power of the base 10, or in other words, 10 raised to some exponent.

the coefficient – one figure before the decimal (3), then any remaining figures (527)

$$3.527 \times 10^3$$

the exponent term – 10 raised to the 3rd power, indicating that the number is multiplied by 10 x 10 x 10

When we convert numbers out of scientific notation, we multiply them by 10 by the number of times expressed in the exponent. In the example given above, the exponent 3 indicates that the coefficient is multiplied by 10 three times. In other words, $3.527 \times 10^3 = 3.527 \times 10 \times 10 \times 10 = 3527$. A rapid conversion can be made by moving the decimal three places to the right (showing that we are multiplying by 10 three times).

$$3.527 \times 10^3 = 3527$$

1 2 3 decimal places to the right

At a quick glance, the value of the exponent tells us a lot about the value of the number. A large exponent indicates a BIG number. $2.3 \times 10^{24} = 2,300,000,000,000,000,000,000,000$!

Sometimes we encounter numbers in scientific notation that contain a negative exponent. Any negative exponent represents a *reciprocal*.

The reciprocal of 10^{-1} is $\dfrac{1}{10^1}$.

The reciprocal of 10^{-3} is $\dfrac{1}{10^3}$ or $\dfrac{1}{10 \times 10 \times 10}$. The reciprocal of 10^{-19} is $\dfrac{1}{10^{19}}$.

These reciprocals indicate that the coefficient is *divided* by 10 by the number of times expressed in the exponent. For instance, 3.9×10^{-3} is converted out of scientific notation by dividing 3.9 by 10 three times. In other words, $3.9 \times 10^{-3} = 3.9 \div 10 \div 10 \div 10 = 0.0039$. A rapid conversion can be made by moving the decimal three places to the left (showing that we are dividing by 10 three times).

$$3.9 \times 10^{-3} = 0.0039$$
3 2 1 decimal places to the left

Again, a quick glance at the exponent reveals details about the number. A negative exponent means that we have a value less than 1.0. The larger the negative exponent the smaller the number.

$$5.9 \times 10^{-15} = 0.0000000000000059$$

Performing Calculations with Scientific Notation

Adding and subtracting

How do we add 2.5×10^2 and 1.0×10^3 when their exponents are different? 10^3 (1,000) is an order of magnitude larger than 10^2 (100). We first need to convert these numbers to the same power of ten. Then we can simply add the coefficients and reiterate the exponent term. In this example we can convert the first number to match the power of ten in the second number.

$2.5 \times 10^2 = 0.25 \times 10^3$.   Now we can add the coefficients and write in the exponent.

$0.25 \times 10^3 + 1.0 \times 10^3 = 1.25 \times 10^3$.

Do you need to prove to yourself that this is correct? If so, take the numbers out of scientific notation, add them, and then convert your answer to scientific notation to compare it to our response above.

$0.25 \times 10^3 = 250$ and $1.0 \times 10^3 = 1000$

$250 + 1000 = 1250$, which is 1.25×10^3

Multiplying

When multiplying numbers in scientific notation we multiply the coefficients and add the exponents.

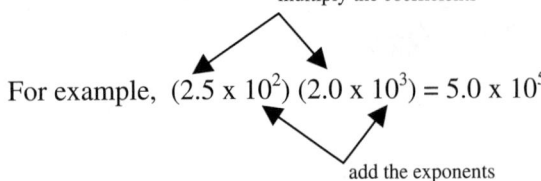

For example, $(2.5 \times 10^2)(2.0 \times 10^3) = 5.0 \times 10^5$.

Here's another example.

$$(6.0 \times 10^2)(2.0 \times 10^3) = 12.0 \times 10^5.$$

But now we have two figures to the left of the decimal when, by convention, we only leave one. We need to change the location of the decimal in our answer so there is only one figure to the left of the decimal.

$$12.0 \times 10^5 = 1.2 \times 10^6$$

Dividing

When dividing numbers in scientific notation we divide the coefficients and subtract the exponents.

For example,

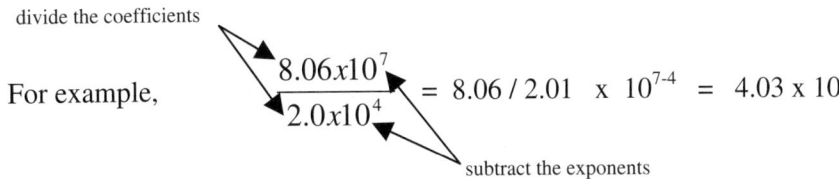

divide the coefficients / subtract the exponents

$$\frac{8.06 \times 10^7}{2.0 \times 10^4} = 8.06/2.01 \times 10^{7-4} = 4.03 \times 10^3$$

Here's another example.

$$3.4 \times 10^9 / 4.0 \times 10^3 = 3.4/4.0 \times 10^{9-3} = 0.85 \times 10^6.$$

But now we have no figures (other than zero) to the left of the decimal when, by convention, we should leave one. We need to change the location of the decimal in our answer so there is one figure to the left of the decimal.

$$0.85 \times 10^6 = 8.5 \times 10^5$$

Exponents raised to an exponent

Sometimes we will encounter numbers expressed in scientific notation that are being raised to a power. In these instances, we raise the coefficient to the indicated power, and multiply the exponent term by that power.

For example, $(4.07 \times 10^6)^4 = (4.07)^4 \times 10^{6 \times 4} = 274.3 \times 10^{24} = 2.743 \times 10^{26}$

raise to 4th power / multiply: 6 x 4

Performing Calculations Using Your Calculator

Now that you understand how to perform mathematical functions with numbers in scientific notation it's time to see if you can do them on your calculator. Use the numeric keypad to input the coefficient, then push the exponent button (labelled EXP, Exp or EE on different calculators) and input the value of the exponent. The +/- button will allow you to put in a negative value (punch it in before the exponent number).

5

Different calculators display the same values in different formats. For instance, 9.87×10^6 may be shown on some calculators as 9.87E6 or as 9.87 x 10^6. All these displays mean the same thing.

It's Your Turn to Flex Some Mental Muscles!

Try these questions. The answers are given below, but no cheating! Try them on your own first, then on your calculator. Work out your own responses before checking them against the answers provided.

1. 3,872,000 =
2. $9.82 \times 10^6 =$
3. 9134 =
4. $1.574 \times 10^3 =$
5. 0.127 =
6. $2.751 \times 10^{-4} =$
7. $3.5 \times 10^4 + 2.2 \times 10^3 =$
8. $3.5 \times 10^4 - 3.9 \times 10^3 =$
9. $(1.9 \times 10^3)(3.2 \times 10^1) =$ 1900 × 32 = 60800 = 6.08×10⁴
10. $(5.2 \times 10^{-5})(1.1 \times 10^4) =$
11. $(8.7 \times 10^6) / (4.3 \times 10^3) =$
12. $(9.5 \times 10^{21}) / (3.2 \times 10^{19}) =$
13. $(3.22 \times 10^{15})^3 =$
14. $(5.67 \times 10^{-8})^4 =$

Answers (shown inverted on the page):

1. 3.872×10^6
2. 9,820,000
3. 9.134×10^3
4. 1.574
5. 1.27×10^{-1}
6. 0.0002751
7. $3.5 \times 10^4 + 0.22 \times 10^4 = 3.72 \times 10^4$
8. $3.5 \times 10^4 - 0.39 \times 10^4 = 3.11 \times 10^4$
9. $(1.9)(3.2) \times 10^{3+1} = 6.08 \times 10^4$
10. $(5.2)(1.1) \times 10^{-5+4} = 5.72 \times 10^{-1}$
11. $(8.7)/(4.3) \times 10^{6-3} = 2.02326 \times 10^3$
12. $(9.5)/(3.2) \times 10^{21-19} = 2.97 \times 10^2$
13. $(3.22)^3 \times 10^{45} = 3.339 \times 10^{46}$
14. $(5.67)^4 \times 10^{-32} = 1.03 \times 10^{-29}$

You might be wondering how many figures to leave in your answers. Where should they be rounded off? We'll tackle that issue when we discuss 'Significant Figures' in Unit 4.

The SI System of Measurement

(Système International d'Unités—International System of Units)

The Système International (SI System, sometimes called the Metric System) is based on seven fundamental units of measure, four of which are especially important to geographers:

i. the **metre** as the unit of length (m),
ii. the **kilogram** as the unit of mass (kg),
iii. the **second** as the unit of time (s), and
iv. the **kelvin** as the unit of temperature (K).

Several other measures, such as the **pascal** as the unit of pressure, the **joule** as the unit of energy, the **watt** as the unit of power and the **newton** as the unit of force, are derived from the seven base units.

Designated prefixes

The International Bureau of Weights and Measures has adopted a series of prefixes and symbols for decimal multiples and submultiples of SI units ranging from 10^{24} to 10^{-24}. These prefixes and symbols are presented below in Table 1.1. Many of them will look familiar. Others, such as the yotta or yocto, may not.

Table 1.1   SI Unit Prefixes and Symbols

Multiple	Prefix Name	Symbol	Submultiples	Prefix Name	Symbol
10^{24}	yotta	Y	10^{-1}	deci	d
10^{21}	zetta	Z	10^{-2}	centi	c
10^{18}	exa	E	10^{-3}	milli	m
10^{15}	peta	P	10^{-6}	micro	µ
10^{12}	tera	T	10^{-9}	nano	n
10^{9}	giga	G	10^{-12}	pico	p
10^{6}	mega	M	10^{-15}	femto	f
10^{3}	kilo	k	10^{-18}	atto	a
10^{2}	hecto	h	10^{-21}	zepto	z
10^{1}	deca	da	10^{-24}	yocto	y

Unit conversions: method #1

Geography students typically need to work with distances measured in millimetres, centimetres, metres, and kilometres. Often it is necessary to convert from one unit to another. For people not familiar with SI unit prefixes and their values, an examination of the following set of prefix relationships may provide clarity. The example given here happens to involve distance in metres, but the prefix relationships hold for any SI unit of measure.

0.001	kilometres
0.01	hectometres
0.1	decametres
1.0	**metres**
10	decimetres
100	centimetres
1,000	millimetres

A metre contains 10 decimetres, 100 centimetres, and 1,000 millimetres. A metre is 1/10th of a decametre, 1/100th of a hectometre and 1/1,000th of a kilometre. How many metres are in a kilometre, then? One thousand metres.

Because the SI system is based on multiples and submultiples of 10 it is easy to convert between units; just follow changes in the location of the decimal and note how many places to the left or right it moves between given units. For example, if we were asked to express millimetres in metres, how many places should the decimal move, and in which direction? Looking first at the location of the decimal in the millimetres line, in what direction would it have to move to convert to metres? How many decimal places must it move?

7

1.0   **metres**
10   decimetres
100   centimetres
1,000.   millimetres

initial decimal is here

In a conversion from millimetres to metres, the decimal must move to the left three places.

1,000 millimetres = 1.0 metre
3 2 1 decimal places to the left

If we measured a distance on a map and found it to be 2,500 metres, how many kilometres would that be?

0.001   kilometres
0.01   hectometres
0.1   decametres
1.0   **metres**

The decimal again moves three places to the left. Applying this to a value of 2,500 m gives:

2,500 metres = 2.5 kilometres
3 2 1 decimal places to the left

If you were given a value of 0.187 km and asked to convert it to metres, what would you do? Can you employ the same strategy? Yes! How many places should you move the decimal, and in what direction?

1 2 3 decimal places to the right

0.001   kilometres
0.01   hectometres
0.1   decametres
1.0   **metres**

From km to m, the decimal moves three places to the right. So your value of 0.187 km = 187 m.

Unit conversions: method #2

When you need to show unit conversions explicitly, a second method may be employed. This can be helpful when you want to show all your workings on a lab exam or when more complex conversions are called for. Let's say, for example, that you need to convert milligrams to kilograms, specifically 984,000 milligrams to kilograms. With this method, you can write out the value you need to convert and then multiple it by a series of ratios that provide the conversions. In the following example the brackets imply multiplication.

$$984,000 mg \left(\frac{1g}{1000mg} \right) \left(\frac{1kg}{1000g} \right) = ?$$

These ratios represent the ratios of milligrams to grams and of grams to kilograms, respectively. By placing milligrams as the denominator of the first ratio and grams as the denominator of the second, we have ensured that when we multiply the terms of this equation the milligrams and grams will cancel, leaving us with an answer in kilograms.

$$984,000 mg \left(\frac{1 g}{1000 mg}\right)\left(\frac{1 kg}{1000 g}\right) = ?$$

milligrams cancel; grams cancel

Now the final mathematical calculations are performed. Three zeros will cancel, leaving us to divide 984 by 1000. Dividing by 10 is the same as moving the decimal to the left by one place, dividing by 100 is the same as moving it two places to the left, and dividing by 1000 is the same as moving the decimal three places to the left. This leaves us with an answer of 0.984 kg.

$$= 984,000 \left(\frac{1}{1000}\right)\left(\frac{1 kg}{1000}\right) = 0.984 kg$$

zeros cancel; move decimal three places to the left to ÷ by 1000

We could leave our answer as 0.984 kg, or if we were asked to express it in scientific notation, we would leave it as 9.84×10^{-1} kg.

Your Turn!

Try the following questions. Work out your responses before checking them against the accompanying answers.

1. 324 m = _____ km
2. 324,000 cm = _____ km
3. 100 g = _____ kg (grams to kilograms)
4. 30 g = _____ mg (grams to milligrams)
5. 2.4×10^6 m = _____ km
6. 842,000 W = _____ MW (Watts to Megawatts)

324 m = 0.324 km
324,000 cm = 3.24 km
100 g = 0.1 kg
30 g = 30,000 mg
2.4×10^6 m = 2.4×10^3 km
842,000 W = 0.842 MW

An Introduction to Maps: A Supplement to Unit 3

Because geographers communicate through both written and spoken language and through numerical and graphical means, learning the art and science of geography requires fluency in the rich language of maps and their symbols. Maps are important in all branches of physical geography and whether you are studying weather and climate, biogeography, and/or geomorphology, you will benefit from a strong working knowledge of maps.

This supplementary unit is designed to introduce Canadian topographic maps and to specifically examine two fundamental elements of all maps: scale and the spherical coordinate system of latitude and longitude.

Canadian Topographic Maps

Transverse Mercator projections

No two-dimensional paper map can portray the exact reality of our three-dimensional planet. As described in Unit 3 of *Physical Geography: the Global Environment*, Canadian edition, trying to create a map of the Earth's surface is like trying to peel the skin off an orange in a single piece which can then be laid flat to form a rectangle. Ideally, the resulting rectangle would portray Earth's spherical surface with no tearing or distortion. But this cannot be done. Every flat map of the earth, likewise, contains some distortion. The topographic maps portraying the physical and cultural features of Earth's land surface that you will examine in geography are produced with great care to minimize such distortion. The map-making process begins with the selection of an appropriate **projection**.

The word 'projection' refers to ways of transferring geographic information from the spherical Earth to a planar map sheet. There are two broad categories of projections: mathematical and true. While most maps created today involve mathematical transformations of our spherical planet, we will examine true projections to give you the conceptual framework with which to understand Canadian topographic maps.

True projections involve the transfer or *projection* of information from the globe onto some screen or projection surface. Picture the globe as clear glass with a light bulb at its centre and with the continents drawn in as thin black lines on its surface. Now picture different ways that we could use the light bulb to accurately project the positions of the coastlines onto sheets of paper. Three different projections could accomplish this, as illustrated in Unit 3 of *Physical Geography*.

1. A plain paper surface could be laid against the glass globe and the coastlines could be projected and drawn onto the paper. There would be no distortion at the point of contact, but distortion would increase with distance from that point. This process creates a **planar projection** that is typically used to map the poles.

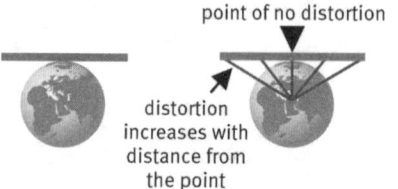
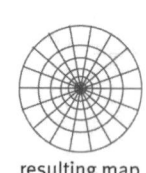

2. A cylinder of paper could be wrapped around the globe, and after the coastlines are projected and drawn onto the paper, the cylinder could be cut open to form a rectangular map. Instead of a point of no distortion, in this **cylindrical projection** a line of no distortion would be formed where the globe and the paper intersect.

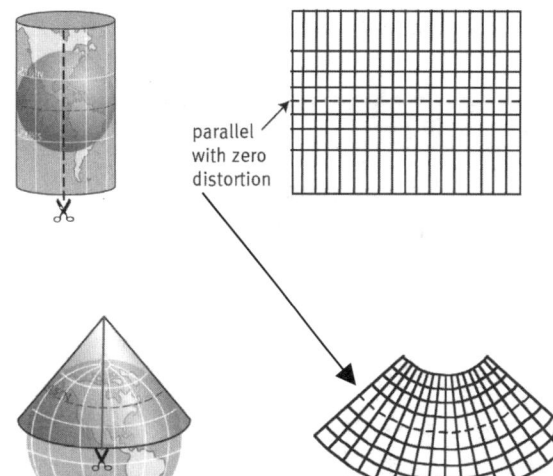

3. A cone of paper could be wrapped around the globe. Would it intersect with the globe at a single point or along a line?

   In this **conic projection** a *line* of no distortion would be formed. When cut open, the line would form an arc.

Which kind of projection—planar, cylindrical, or conic—would be best for Canadian maps? A planar projection would show the Arctic with little distortion, but not other parts of the nation. A conic projection might show the middle latitudes with accuracy, but not the Canadian north.

To map Canada with great precision a cylindrical projection called the **Mercator projection** is used in a very specific way.

If we wrapped a cylinder of paper around the globe such that the north and south poles were aligned with the open ends of the paper cylinder, would the line of no distortion run from pole to pole or would it travel around the equator?

Let's think this through. A cylinder is wrapped around the globe with the open ends aligned with the poles.

 The line of contact between the globe and the paper cylinder will be along the equator. This will not produce a line of zero distortion in Canada.

What would happen if we changed the orientation of the cylinder by 90°? This is said to be a **transverse** orientation.

With a transverse orientation the open ends of the cylinder are centred at the equator.

 The line of zero distortion runs from N–S along a meridian of longitude.

If we only mapped near that line of zero distortion, this could produce an accurate map of a N–S slice of Canada. And indeed that's how our mapping is done: in thin slices called **zones** produced from transverse Mercator projections. Each zone extends only 3° of longitude on either side of the central meridian to minimize distortion. Zones run from 84° North Latitude to 80° South Latitude (Figure 3.1). It takes 60 of these longitudinally 6°-wide zones to span our 360° planet. The strips that cover Canada include zones 7

11

to 22, as indicated in Figure 3.2. These transverse Mercator zones leave out the polar regions, which are mapped with a planar projection.

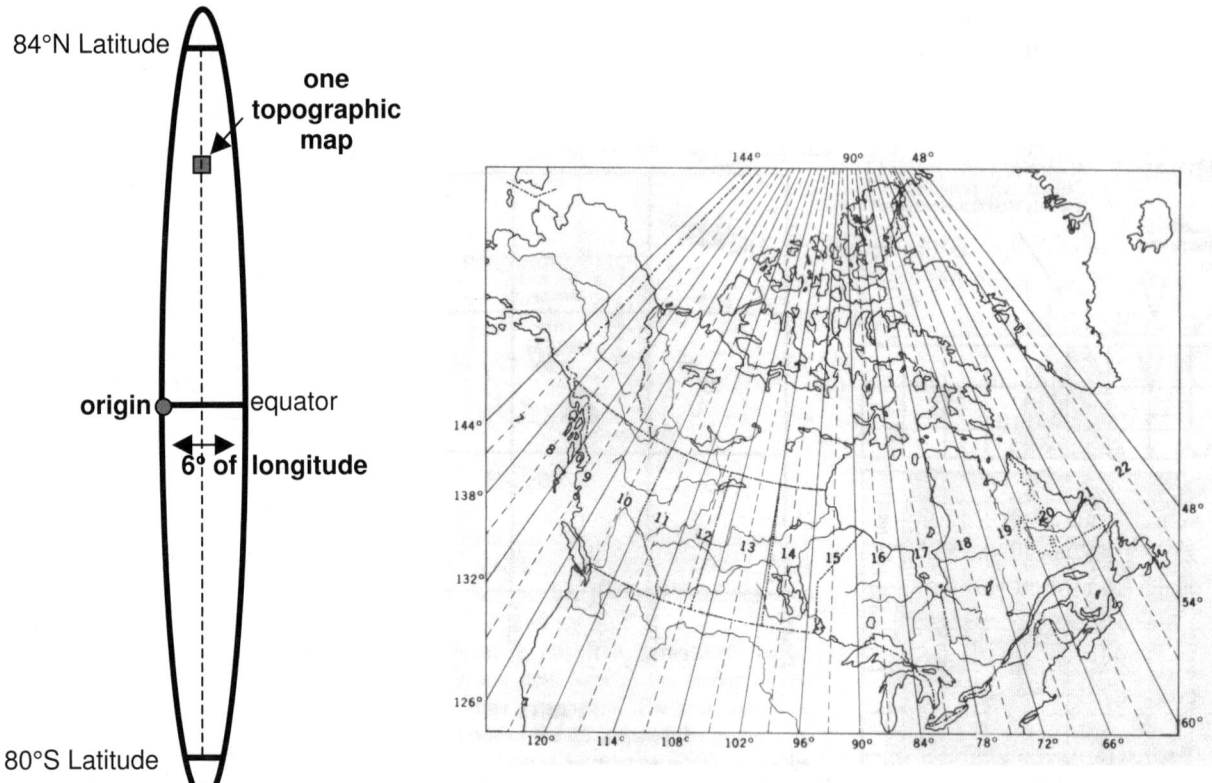

Figure 3.1 Transverse Mercator projection zones. The UTM grid originates at the equator and the western edge of the zone.

Figure 3.2 UTM Zones and Central Meridians for Canada. Source: http://maps.nrcan.gc.ca/topo101/utm2_e.php. © Topographic map reproduced under licence from Her Majesty the Queen in Right of Canada, with permission of Natural Resources Canada.

A more complete examination of topographic maps will begin in Section A of this workbook, but for now we'll focus on two map elements that will be helpful for studies in any aspect of physical geography: scale and latitude/longitude.

Map Scale

A map can be defined as a representation to scale of features on a real surface. While we might have a sense of what the word *scale* means, we need a formal definition that will allow us to measure and analyze features on maps.

As reported in Unit 3 of *Physical Geography*, scale is the ratio of map distance to the ground distance it represents. The scale is sometimes written as a fraction, for instance:

$$\frac{1}{25,000} \quad \text{or} \quad \frac{1}{200,000}$$

Typically it is provided as a ratio or **representative fraction** of, in this case, 1:25,000 or 1:200,000. You may notice such scales are *unit-less*—they have no units. This is the beauty of a representative fraction scale; you can choose the units you wish to work with. All you have to do is maintain the ratio of map to ground distance.

Representative Fraction (RF) Scale

Remember that the scale is the ratio of map distance to the ground distance it represents.

1:25,000

This means that one of some unit on the map represents 25,000 *of the same units* on the ground. The first number in the ratio (or the numerator if the scale is presented as a fraction) is always one. Because it's often convenient to measure map distances in centimetres, we'll work through an example using those units.

Let's say that you were asked to measure the straight-line distance between two points on a map. Using your ruler, you identified the map distance as 10 cm. The scale of the map is 1:25,000, so every centimetre on the map represents an actual ground distance of 25,000 centimetres. If 1 cm map distance = 25,000 cm ground distance, what does your distance measurement of 10 cm on the map represent? It shows that the actual ground distance between the features must be (25,000 cm)(10) = 250,000 cm. Remembering what you've learned about SI units, you could convert that awkward value (250,000 cm) to metres or kilometres. Moving two decimal places to the left to convert from centimetres to metres would give a value of 2,500 m. Moving a further three decimal places to convert from metres to kilometres would give a value of 2.5 km.

Working through the math

You could figure out the previous example in your head. But if you measured a map distance of 7.63 cm, you would need a strategy to determine the ground distance mathematically. If you have a strong science background you will already know how to create the following equations, so feel free to skip ahead. If your science background is more limited, this section is designed to help you develop some methods to deal with 'word problems' that often crop up in geography.

You've been asked to calculate the straight-line distance between two points on a map with a scale of 1:25,000. You've measured the distance on the map as 7.63 cm. Now you have to determine what that distance is on the earth's surface. How will you do that?

You could use a 'ratio of what you know' to a ratio of 'what you need to find out' to solve for an answer.

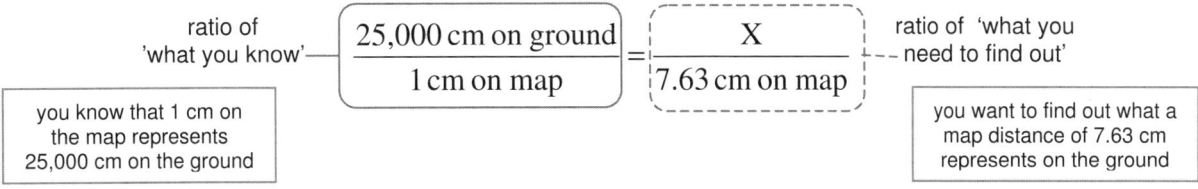

An equation like this allows you to solve for X. How could you isolate X on the right side of the equation?

$$\frac{25{,}000 \text{ cm on ground}}{1 \text{ cm on map}} = \frac{X}{7.63 \text{ cm on map}}$$

If you multiplied the right side by 7.63cm, that would cancel out the denominator.

$$\frac{25{,}000 \text{ cm on ground}}{1 \text{ cm on map}} = \frac{X \;\;\cancel{(7.63\text{ cm})}}{\cancel{7.63 \text{ cm on map}}}$$

To preserve the equality of the equation, you would also have to multiply the left side by 7.63 cm. Aha! You've isolated X and now can finish solving the equation. Notice that the shortcut version to solve this formula would be 'cross-multiply and divide'.

$$\frac{(7.63\text{cm})(25{,}000\text{ cm})}{1\text{cm}} = X$$

Solving for X would produce 190,750 cm. Again, moving two decimal places to convert to metres and another three to convert to kilometres would give 1.9075 km. You could also show your conversions explicitly with:

$$190{,}750 cm \left(\frac{1m}{100cm}\right)\left(\frac{1km}{1000m}\right) = 1.9075 km$$

We'll examine how to round off a final answer in Unit 4.

Different Expressions of Scale

Scales may be expressed not only as a representative fraction, but also as a **statement in words** and as a **graphic** or **bar scale**. A statement in words for the 1:25,000 scale we examined above would be, *'one centimetre represents two hundred and fifty metres'*. Note that we say 'represents' rather than 'equals' because, clearly, 1 cm is not equal to 250 m. A graphic or bar scale is given as a length of line divided into numbered segments scaled to represent the cited ground distances.

You could measure a map distance with a string or piece of paper and compare it to the bar scale. This method, however, is awkward for large distances and is less precise and accurate than working with a representative fraction scale.

Large scale and small scale maps

When geographers refer to large scale and small scale maps they are not talking about the outer dimensions of the maps, they are referring to the representative fraction scales. Scales of 1:25,000 and 1:200,000 expressed as fractions are, respectively:

$$\frac{1}{25,000} \quad \text{and} \quad \frac{1}{200,000}$$

Which is the larger fraction? One twenty-five thousandth is much larger than one two hundred thousandth, so 1:25,000 is a larger scale than 1:200,000.

Examine a wall map of the world in your classroom. Its scale will probably be on the order of 1:30,000,000—a very small scale! What generalizations can you make from observations of large and small scale maps? Mentally compare a classroom world map at a scale of, say, 1:10,000,000 and a road map for your city at a scale of 1:20,000. Which would show a larger area of the earth's surface? Which would show a more detailed view? Large scale maps show a smaller portion of the surface in much greater detail than do small scale maps.

Let's examine the two maps with identical outer dimensions given below in Figure 3.3. One has a scale of 1:50,000. The other has a scale of 1:25,000. Which is larger? The 1:25,000 scale is two times larger than the 1:50,000 scale. Notice the relationship between the scale and map area expressed in this figure. If the outer dimensions are identical, a 1:25,000 scale map will depict ¼ the area of a 1:50,000 map. How could that be if the scale only differs by a factor of two?

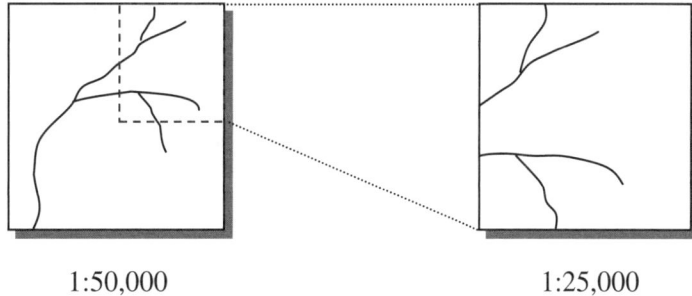

1:50,000   1:25,000

Figure 3.3 The relationship between map area and scale. If the outer dimensions are the same, the difference in area shown by two maps of different scale will be the square of the difference in map scale.

Remember that area is equal to length multiplied by width. $A = (l)(w)$. If the 1:25,000 map shows half the distance across the length of the map and half the distance across the width of the map, then the difference in area is:

$$\frac{1}{2} \times \frac{1}{2} = \frac{1}{4}$$

Another way of looking at this is to examine the difference between the two representative fraction scales.

$$\frac{1}{25,000} \times \boxed{\frac{1}{2}} = \frac{1}{50,000}$$

difference in scale

15

We can generalize from this to say that the difference in area will be equal to the square of the difference in scale.

$$\left(\frac{1}{2}\right)^2 = \frac{1}{4}$$

> **Some Questions for You!**
>
> Working with scale is an essential skill for any geographer. Try these questions out to test your understanding and working knowledge.
>
> 1. Given a map scale of 1:25,000, what ground distance is represented by:
>   a. 0.5 mm on the map, and
>   b. 2 cm on the map?
> 2. How many centimetres on this map scale represent 1 km of ground distance?
> 3. How would you express this scale (1:25,000) as a statement in words?
> 4. Which is larger, a 1:20,000 scale or a 1:100,000 scale?
> 5. What is the difference in area covered by two maps of the same dimensions, one at a scale of 1:20,000 and the other at a scale of 1:100,000? Which map will show the larger area?
>
> The answers are given on the last page of this unit.

Locational Systems on NTS Map Sheets

Two systems for communicating location are utilized on Canadian NTS maps:

1. the spherical coordinate system of latitude and longitude examined below, and
2. the Universal Transverse Mercator grid, which is examined later in Section A.

The spherical coordinate system: latitude and longitude

Geographers need to understand how to cite and calculate values of latitude and longitude. In addition to dealing with the location of landforms or places, geographers must determine measures such as the intensity of solar radiation received at a given place and time, which is dependent on the latitude of the place.

The spherical coordinate system was introduced in Unit 3 of the textbook. You will remember that the north–south dimension of this system is called latitude. Lines of latitude are often called **parallels** because they are parallel to the equator. They indicate how far north or south a location is from 0° latitude (the equator, a natural starting place that is halfway between the north and south poles). The east–west dimension of the spherical coordinate system is called longitude, and is expressed by N–S running lines named **meridians**. Meridians specify distance east or west of the Prime Meridian (0° longitude, an arbitrarily appointed starting place set at an international conference in 1884).

Latitude and longitude values symbolize measures of arc from the centre of the earth. Even a small 1° difference of latitude when extended to the surface of the planet translates into a distance of more than 100 km. Citing a ballpark location that could be 100 km away from the actual location is not acceptable precision! For this reason degrees are subdivided into 60 smaller units called *minutae primae* in Latin, meaning 'the first small parts', that we call **minutes**. Degrees are indicated with ° (degree symbol) and minutes are denoted with ′ (minute symbol, which resembles a straight comma). Each minute of latitude still represents a distance of more than a kilometre, so minutes are subdivided into 60 smaller parts called *minutae secondae* in Latin, meaning 'the second small parts' that we call **seconds**. Seconds are represented by a symbol that looks like a double quote (″). Note that the words 'minutes' and 'seconds' used in this sense do not describe segments of time. In this connotation they represent subdivisions of a degree; they are measures of arc. In summary, 1° = 60′ and 1′ = 60″.

The locations of three Canadian cities are given below.

City, Province/Territory	Latitude and Longitude Coordinates
Halifax, Nova Scotia	44° 38′ N, 63° 30′ W
Iqaluit, Nunavut	63° 45′ N, 68° 33′ W
New Westminster, British Columbia	49° 12′ N, 122° 54′ W

We can express the coordinates for Halifax in words. Halifax is located at forty-four degrees and thirty-eight minutes north latitude, sixty-three degrees and thirty minutes west longitude. Two important conventions are visible in these citations. Firstly, latitude is always given before longitude. Secondly, with few exceptions, a north or south direction must be designated for latitude and an east or west direction must be communicated for longitude.

There are three exceptions to this rule. Because there is only one 0° latitude, no north or south designation is necessary when citing the equator. Similarly, there is only one 0° and one 180° longitude, so no east or west designation is necessary when citing either the prime meridian or the international dateline. For every other parallel or meridian a direction designation must be given.

Performing Calculations with Latitude and Longitude

In climatology we might want to calculate the angle of the noon sun above a specific location in order to estimate the intensity of solar energy received. On a less scientific front, we might one day want to calculate the angle of the noon sun to see whether a house we're considering purchasing will be entirely shaded by the neighbouring buildings. In both instances we would need to be able to add and subtract angles of arc.

Performing simple calculations with angles of arc may not seem straightforward at first. Would you know how to subtract 2°58′ from 23°17′? Plugging 23.17 - 2.58 into our calculators would be incorrect because 17 minutes of arc is 17/60th of a degree, not 0.17 or 17/100th of a degree.

If you have experience with these kinds of computations, skip ahead. If these ideas are new, this section is written for you.

How can we subtract 2°58′ from 23°17′? We'll need to deal with the minutes as a column of numbers and with the degrees as another column. We can't subtract 58′ from 17′ without borrowing a degree. If we borrow one degree, we are left with 22°. Because there are 60′ in every degree, and because we already had 17′, we now have 77′. We can re-express 23°17′ as 22°77′ to allow this subtraction. Now we can subtract 58′ from 77′ and get 19′. Similarly, 22° - 2° is 20°. So our final answer is 20°19′.

```
 23° 17′    =     22° 77′
- 2° 58′         - 2° 58′
                  20° 19′
```

Some people have calculators that will perform computations with measures of arc. Those people should be aware of two potential problems. Firstly, it is crucial to be able to reason through a quantitative problem in order to know what to do with your calculator. Secondly, you must ensure that your calculator is in *degree* mode or your answers will all be incorrect.

Some Questions on Latitude and Longitude

Try these questions out, and then check your answers against those given at the end of this unit.

1. Lines of latitude are also known as _____ while lines of longitude are known as _____.
2. Lines of latitude run ____ (N–S/E–W) and represent distance _____ (north or south / east or west) of the equator.
3. Lines of longitude run ____ (N–S/E–W) and represent distance ____ (north or south / east or west) of 0° Longitude.
4. Which lines of latitude or longitude do not need a direction designation? Why?
5. When using the spherical coordinate system _____ is always cited before _____.

Performing Calculations with Latitude and Longitude

1. Sum the following latitudinal values:

   47° 42′ 23″   23° 59′ 57″
   + 1° 12′ 12″   + 7° 15′ 06″

---

2. Subtract the following latitudinal values:

   47° 42′ 23″   23° 59′ 06″
   - 1° 12′ 12″   - 7° 15′ 23″

Answers to Scale Questions

1. Given a map scale of 1:25,000, what ground distance is represented by:
   a. 0.5 mm on the map, and
   b. 2 cm on the map?

   Employ the strategy of writing the ratio of 'what you know' to the ratio of 'what you need to find out'. At a scale of 1:25,000, one millimetre on the map represents 25,000 millimetres on the ground.

   $$\frac{25,000mm}{1mm} = \frac{X}{0.5mm}, \text{ so } X = \frac{(0.5mm)(25,000mm)}{1mm} = 12,500mm = 12.5m$$

   Using the same system, at this scale 1 centimetre on the map represents 25,000 cm on the ground.

   $$\frac{25,000cm}{1cm} = \frac{X}{2cm}, \text{ so } X = \frac{(2cm)(25,000cm)}{1cm} = 50,000cm = 500m$$

2. How many centimetres on the map represent 1 km of ground distance? *Try working backwards.*

   $$1km = 1000m = 100,000cm, \text{ so } \frac{1cm}{25,000cm} = \frac{X}{100,000cm}, \text{ so } X = \frac{(1cm)(100,000cm)}{25,000cm} = 4cm$$

3. How would you express this scale (1:25,000) as a statement in words? *We know that one centimetre represents 25,000cm or 250 metres, so a statement in word would be "one centimetre represents two hundred and fifty metres".*

4. Which is larger, a 1:20,000 scale or a 1:100,000 scale? *A twenty-thousandth is much larger than a one hundred thousandth, therefore, a map scale of 1:20,000 is larger than 1:100,000.*

5. What is the difference in area covered by two maps of the same dimensions, one at a scale of 1:20,000 and the other at a scale of 1:100,000? Which map will show the larger area? *The difference in scale is 1/5. The difference in area is equal to the square of the difference in scale, and therefore, is $(1/5)2 = 1/25^{th}$. The map with the scale of 1:100,000 will show twenty-five times more area than the map at a 1:20,000 scale.*

Answers to Latitude/Longitude Questions:

1. Lines of latitude are also known as *parallels* while lines of longitude are known as *meridians*.

2. Lines of latitude run *E–W* and represent distance *north or south* of the equator.

3. Lines of longitude run *N–S* and represent distance *east or west* of 0° Longitude.

4. Which lines of latitude or longitude do not need a direction designation? Why? *Only meridians and parallels that are not repeated in each hemisphere do not need a direction designation because they cannot be confused. In contrast, 49° latitude, for instance, occurs in both northern and southern hemispheres, so must be specified.*

5. When using the spherical coordinate system *latitude* is always cited before *longitude*.

Answers to Calculations with Latitude and Longitude

1. Sum the following latitudinal values:

       47° 42′ 23″   23° 59′ 57″
   + 1° 12′ 12″   + 7° 15′ 06″
       48° 54′ 35″   30° 74′ 63″ = 31° 15′ 03″

   (60″ so bumps up minutes and degrees)

2. Subtract the following latitudinal values:

       47° 42′ 23″   23° 59′ 06″ = 23° 58′ 66″
   - 1° 12′ 12″   - 7° 15′ 23″   - 7° 15′ 23″
       46° 30′ 11″   16° 43′ 43″

How did you do? Have you mastered these concepts and calculations, or is more review in order?

Significant Figures and Graphs: A Supplement to Unit 4

In the previous sections the numerical answers have been calculated but not rounded off. The question of where to round off numbers, it was explained, would be tackled in Unit 4. And so it shall! This section of the supplement examines:

1. significant figures, and
2. the development and interpretation of graphs.

Significant Figures

Imagine that it's the first geography lab of the year and your instructor has asked you and two of your classmates to take thermometers outside to measure the air temperature. You record a value of 22.3°C, but your other two classmates record temperatures of 22.4°C and 22.1°C, respectively. Because there's small variation in how people read instruments, and because there could be slight calibration differences among thermometers, it's not a surprise that the values are different. A common method to reduce potential measurement errors is to take multiple measurements and then to average the values.

When you and your classmates calculate the average of your temperatures you get 22.2666666667°C. What temperature you should share with the class? Where should you round the number off? If you leave all those decimals you are implying that you have confidence in this temperature measurement down to tenths, hundredths, thousandths. . .to less than a *billionth* of one degree Celsius. Happily, there is a way out of your dilemma. Scientists have already run into this problem and have worked out a method for deciding how many figures to leave.

The custom is to include all figures that are reliably known, plus one more that is an estimate, or approximation (the first uncertain figure). In your temperature reading you know that the temperature is 22°C, so you would include those two figures, plus one more—the first uncertain figure. Your final temperature, then, should be rounded to three digits: 22.3°C. These three digits are said to be **significant figures**.

Determining the number of significant figures in a number is not always easy. Zeros pose special challenges because sometimes it's not obvious whether they are indicating a value or are merely showing the location of the decimal point. Again, some conventions have been worked out to communicate when zeros are significant.

Rules for Zeros

All non-zero digits are significant.

There are four significant figures in 9,273 and also in 3.284. There are three significant figures in 461 and 7.99.

Zeros that simply fix the decimal point are not significant.

There are three significant figures in 24,700. 'Why is that, when five digits are involved?' you may wonder. The number 24,700 tells us that we know this value reliably to twenty-four thousand. The first uncertainty creeps in at the seven hundred mark. We do not have enough resolution to know the value down to the 10s level or lower with any reliability. So the two zeros are not significant. They are important in telling us about the magnitude of the number (it is two hundred and forty-seven *thousand*, not two hundred and forty-seven), but they are not significant. In the same way, the number 0.000523 also has three significant figures. The zeros just 'fix' the decimal point. In other words, the zeros tell us the order of magnitude.

Zeros within a number are always significant.

There are four significant figures in 2,047 and in 0.09105. In the latter number the four significant digits are 9105. The previous two zeros simply fix the decimal place.

Trailing zeros to the right of the decimal point are always significant.

The number 5.1920 has five significant figures. The zero is not needed to tell us where the decimal is, therefore, the only reason it would have been included is to indicate significance down to the 1/10,000th level. The number 67.0000 has six significant figures because, again, these zeros are not needed to fix the decimal. They are significant.

Expressing numbers in scientific notation is often helpful to avoid confusion about when zeros are significant. The number 5,974,000 seems to contain four significant figures. But what if the scientist who measured this value actually knows it reliably down to the hundreds, tens, and ones? If he or she had written it as 5.974000×10^6 then we would know that there are seven significant figures.

Geographers often need to perform calculations and leave their answers to a reasonable number of significant figures. The conventions for multiplying and dividing, and then for rounding off are given below.

Multiplying and Dividing

When multiplying or dividing numbers, the results can only be as accurate as the *least* number of significant figures used in the calculations. For example, if you multiplied 9.84321 (6 sig. figs.) by 230 (2 sig. figs.), initially, you would get 2263.9383. This answer implies a high degree of precision. The data we used in our calculation were not that precise, so it would be dishonest of us to claim that we know the answer to a level of 7 significant figures. Properly expressed in scientific notation, our answer would only have 2 significant figures and would be expressed as 2.3×10^3.

Rounding Off

When you need to reduce the number of figures in an answer, round off using the following rules:

1. If the first 'insignificant' figure (in other words, the first figure you'll drop) is less than 5, drop off all the trailing 'insignificant' figures, and leave the last significant figure unchanged. For example, leaving the following values to two significant figures would change 1.93 to 1.9, change 7.822 to 7.8 and 943.7 to 940.

2. If the first insignificant figure is equal to or greater than five, drop off all the trailing 'insignificant' figures and increase the last significant figure by one. For instance, leaving the following values to two significant figures would change 1.652 to 1.7, 927.2 to 93.0, and 5.85 to 5.9.

Some Questions for You to Try!

1. How many significant figures are in the following numbers?

   a. 300 _____   e. 300.0 _____

   b. 0.009 _____   f. 0.0090 _____

   c. 6590 _____   g. 0.6590 _____

   d. 0.0351 _____   h. 0.3051 _____

2. Tracking a warm front on a series of maps indicates that it is moving at a velocity of 27.42km per 1.0 hours. How far will it go in 3.00 hours?

   Answers are provided at the end of this unit.

Graphs and Graphing

Sometimes the most effective way to visually communicate relationships among data is through graphs. This section of Unit 4 examines:

1. basic components of graphs,
2. key concepts in their interpretation, and
3. some pointers for the development of quality graphs.

Graphs and their components

Graphs are diagrams that show what happens to one or more **variables** as another variable changes. A variable is a measurable item whose value can change or vary, hence the name. Graphs usually contain one horizontal (X) axis, which represents the **independent variable**, with values controlled by the graph maker. Simple graphs contain one vertical (Y) axis, which represents the **dependent variable**, whose values rely on the independent variable. More complex graphs may contain two Y-axes.

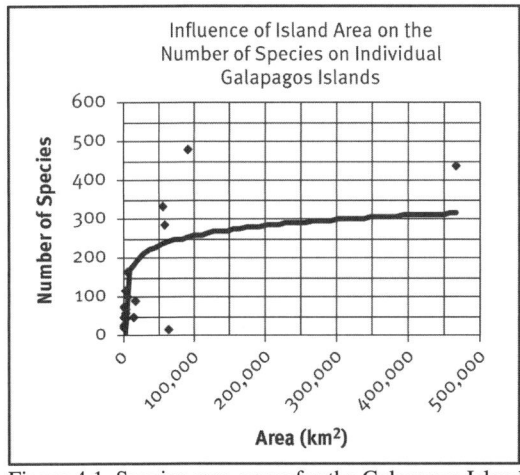

Figure 4.1 Species-area curve for the Galapagos Islands.

For an example, let's examine Figure 4.1 to the left, which expresses an island biogeography relationship known as a species-area curve. We can see that the area of different Galapagos islands has been plotted on the X-axis, so area is the independent variable.

The number of species existing on islands varies with the island size, so the number of species is the dependent variable and is plotted on the Y-axis (the vertical axis).

Like other graphs, Figure 4.1 has labels on both its axes with units given wherever they are not obvious. The X-axis is unambiguously labelled 'Island Area' and the units of measurement are cited as square kilometres (km^2). The Y-axis is unambiguously labelled 'Number of Species'. A descriptive title is given which indicates the variables that are graphed and the geographic location the data were collected in. You should write an informative title for all graphs you create.

Interpreting Graphs

When reading a graph there are several things to watch for. Is the relationship the graph depicts direct or indirect? Is it linear or exponential? Have the axes been plotted to emphasize some point? Can you use the graph to estimate specific values?

Direct relationships
In Figure 4.2 presented to the right, as fictional variable X increases, so too does fictional variable Y. Conversely, as variable X decreases, variable Y does too. The relationship between these two variables is *direct*—as one increases so does the other.

Linear relationships
This example also illustrates the concept of a *linear* relationship, where change occurs in a straight line. We can use linear functions to estimate values of Y given particular values of X. If on Figure 4.2 for instance, X had a value of 9.5, what would Y's value be? We could extend a line from 9.5 units on the X axis up to the function and across to see what the corresponding Y value would be. Here, it would be roughly 190 units.

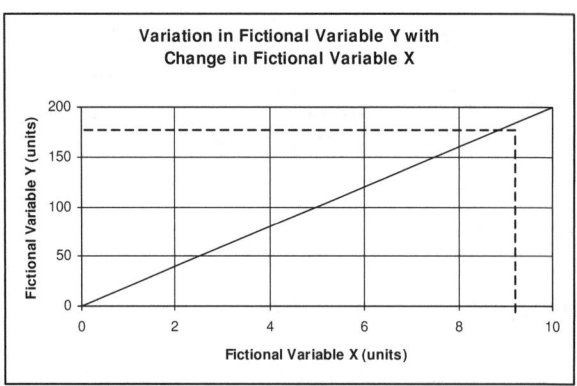

Figure 4.2 Fictional graph illustrating a direct relationship.

Indirect relationships
Figure 4.3 again illustrates a linear relationship, but this time as variable X decreases, variable Y increases. This is an *inverse* relationship; as one variable goes up, the other goes down. This example also shows that, depending on the purpose of the graph and the range of variables to be plotted, the origin of the graph may not be 0, 0. It may be important to show negative values.

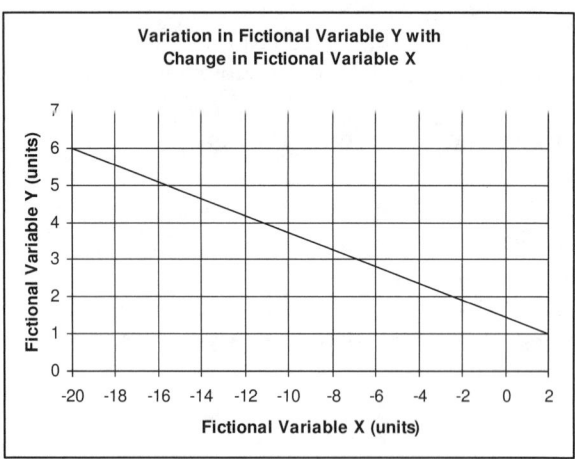

Figure 4.3 Fictional graph illustrating an indirect relationship.

Exponential relationships

Figure 4.4 represents an exponential relationship between variables Y and X. In this case the function it portrays happens to be $Y = 2^X$. The relationship is direct; as one variable increases so does the other, however, the relationship is not linear. The rate of increase is exponential.

Plotting to emphasize

Figures 4.5a and 4.5b below present two very different looking graphs developed from the same data set. Figure 4a presents the data over a 20+ year period and with a vertical scale that visually minimizes the increase in the atmospheric concentration of a fictional gas.

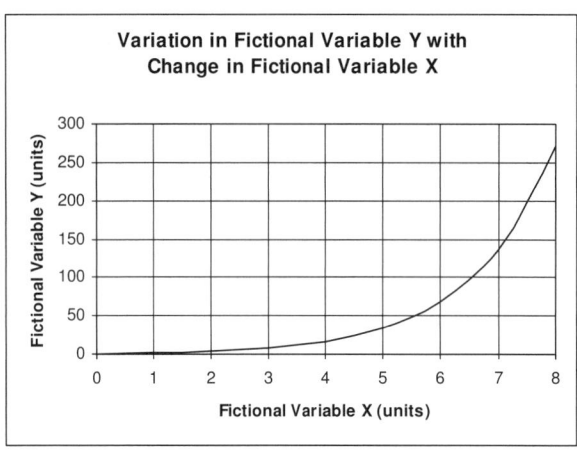

Figure 4.4 Fictional graph showing an exponential relationship.

Figure 4b presents only the last four years of this data set, with a vertical scale designed to emphasize the changes in gas concentration. As a reader of graphs, it is your responsibility to carry out a visual assessment of the scales used to determine the strengths, weaknesses, or possible biases of the picture the graphs present.

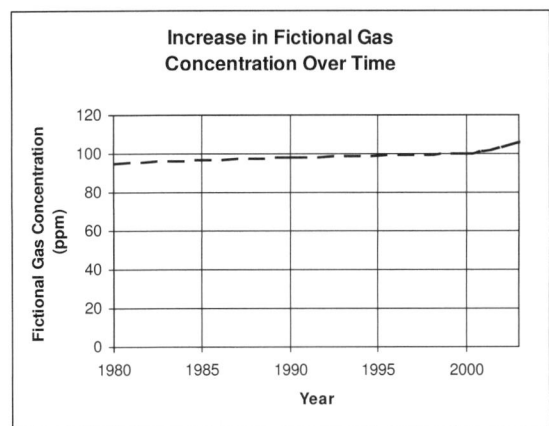
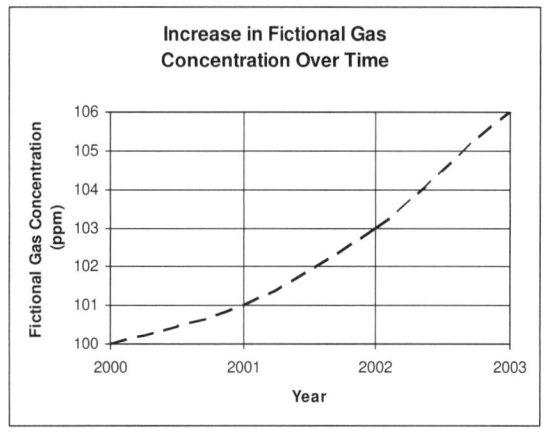

a. b.

Figure 4.5 Two graphs present different pictures of similar data. Figure 4.5a minimizes changes in the atmospheric concentrations of a fictional gas from 1980 to 2003. Figure 4.5b shows the same data, but only from 2000 to 2003. It uses a different vertical scale that emphasizes recent increases in gas concentration.

In the previous section we only examined line graphs, but you should be aware that many different types of graphs exist. Line graphs visually illustrate continuity of change and variation over time, and are best used for those purposes. Bar graphs lend themselves to comparisons of individual quantities. The bars can be oriented horizontal or vertically, in which case they are called columns. Pie charts are often used to compare categories of data.

Some graphs are complex in the sense that they contain more than one type of graph within them. Figure 4.6, for instance, contains a line graph to show temperature in degrees Celsius and a stacked column graph to show precipitation types (white for snow, dark grey for rain) and precipitation amounts in millimetres. Because this graph is part of series, it does not have fully labelled axes and a descriptive title—they are implied by the rest of the series. Any graph you create should have these features.

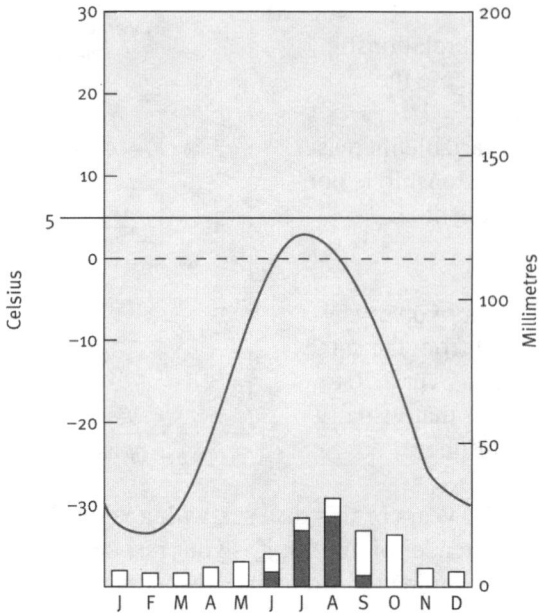
Figure 4.6 Climograph of a high-latitude tundra (**ET**) weather station.
(Fig. 19.8 in H. de Blij, *Physical Geography*, 2005).

Development of Quality Graphs

Your presentation of quantitative data will take on professional qualities as you progress through your studies. If you take the time now to learn to develop quality graphs you will be better prepared for the workforce or for further education.

1. Use a sharp pencil and use a ruler to draw straight, precise lines on your graphs. Print all words and numbers with clarity. If you don't yet know how, over time learn to use a computer to produce professional quality work.

2. Assess the data to be plotted. Determine which type of graph will best suit its attributes.

3. Choose scales that create an honest representation of the data and allow clear plotting and reading of the graph. The X- and Y-axes usually do not have to be plotted to the same scale.

4. Plot the quantity that you choose to vary (the independent variable) on the horizontal axis, and the resulting quantity (the dependent variable) on the vertical axis. An exception to this rule may occur when height or depth is involved because readers will intuitively expect to see them portrayed vertically.

5. Identify the variables plotted along with their units and scales in the following manner:
    a. parallel to but centred beneath the X axis, name the X axis variable and state its units,
    b. number the divisions of the X axis from left to right,
    c. orient the name and units of the Y axis parallel to and centred within the area to the left of the vertical axis, and
    d. number the divisions of the vertical axis from the bottom of the graph upwards.

6. Plot the data with precision and accuracy. If more than one set of data is being plotted on the same graph, use different symbols and/or colours to distinguish the variables. Create a legend in an empty

corner of the graph to decode the symbols/colours for readers. The symbols in the legend must match those in the graph exactly.

7. Centre a full descriptive title above the graph. The title should 'stand alone', in other words, the reader should be able to understand what the graph represents just by reading the title rather than by having to search through the text of the paper for an explanation.

8. If you need to identify your name on the graph before handing it in, place your name on the lower right hand corner of the graph. Be humble; it shouldn't be the largest print on the page.

Figure 4.7 illustrates some of the instructions listed above. Note that the title describes the type of data presented, as well as the location and time the data was collected. The horizontal and vertical axes are clearly labelled with the names and units of the variables they represent. A simple legend placed in a corner decodes the symbols of the graph.

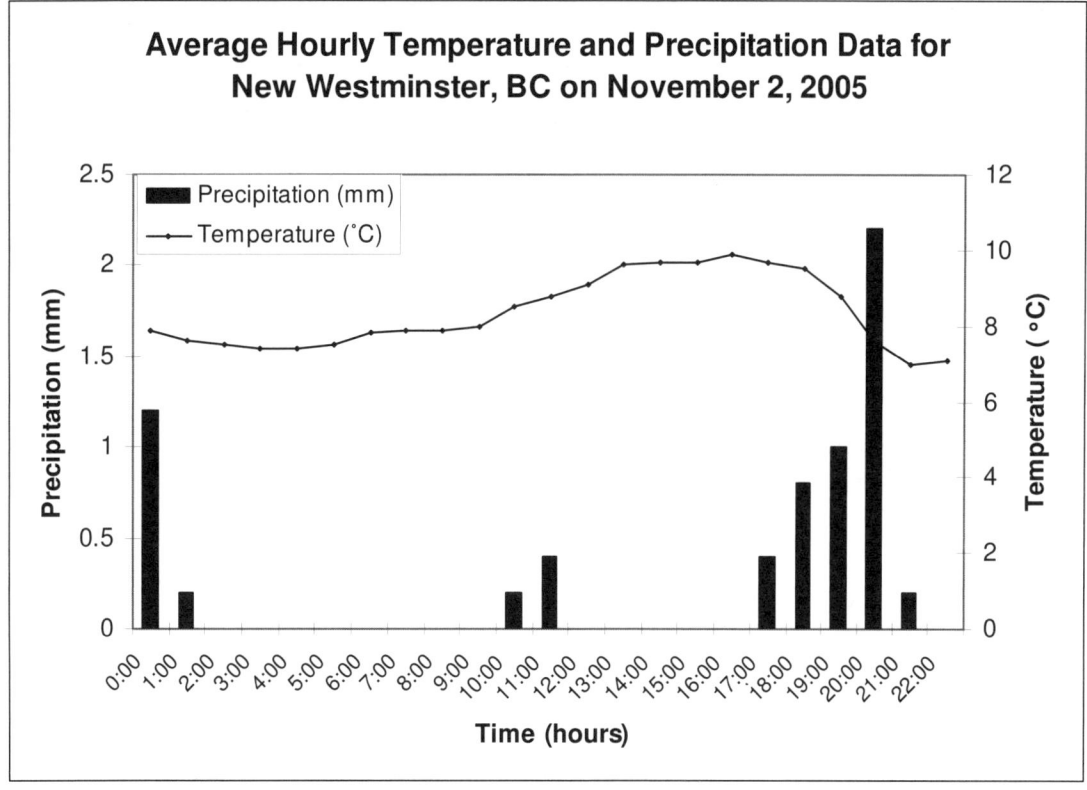

Figure 4.7. Average hourly temperature precipitation measurements for New Westminster, British Columbia, November 2nd, 2005.

Answers to the Significant Figures Questions

1. How many significant figures are in the following numbers?

    e. 300   *one sig. fig. – 3*   e. 300.0   *four sig. figs. – 3, 0, 0, 0*

    f. 0.009   *one sig. fig. – 9*   f. 0.0090   *two sig. figs. – 9, 0*

    g. 6590   *three sig. figs. – 6, 5, 9*   g. 0.6590   *four sig. figs. – 6, 5, 9, 0*

    h. 0.0351   *three sig. figs. – 3, 5, 1*   h. 0.3051   *four sig. figs. – 3, 0, 5, 1*

2. Tracking a warm front on a series of maps indicates that it is moving at a velocity of 27.42 km per 1.0 hours. How far will it go in 3.00 hours?

    This example was made up to illustrate a rounding method; note that we can't track an air mass with this level of precision! We need to set up an equation to solve this word problem. We can do that using the method presented for scale calculations in Supplement Unit 3. We know that the front travels 27.42 km in 1.0 hour. This is the 'ratio of what we know'. We want to find out how far it will travel in 3.00 hours.

    $$\frac{27.42 km}{1.0 h} = \frac{X}{3.00 h} \quad \textit{Isolating for X gives} \quad X = \frac{(27.42 km)(3.00 h)}{1.0 h} = 82.26 km, \textit{which must be rounded}$$

    off to the least number of significant figures in our initial data. The least number is two significant figures (1.0 h), so our final answer should be 82km/h.

Earth–Sun Relationships: A Supplement to Unit 5

Anyone who has been outside during the Canadian winter can attest to the fact that the sun is relatively low in the sky, and that when the sun is in that position it does not produce appreciable surface warming. Snowbirds who escape the cold winter temperatures understand that latitude plays a large role in influencing sun angles, which in turn have great impact on solar heating of the surface.

Geography laboratory assignments sometimes explore these Earth–Sun relationships by examining solar altitudes and the intensity of solar energy received at given times and latitudes. This segment of the supplement is designed to provide further background, sample calculations, and practice questions on the following concepts:

1. the subsolar point and solar declination, and
2. the solar altitude or elevation.

Subsolar Point and Solar Declination

The **subsolar point** is the place on the earth's surface where the sun is directly overhead; it is perpendicular to the surface. The **solar declination** is the latitude of the subsolar point at local solar noon on a given day. As illustrated in Figure 5.2 from the text, the solar declination is at 23 ½ °N at the summer solstice on June 22nd and at 23 ½ °S at the winter solstice on December 22nd. At all other times of the year it varies between these latitudes. The solar declination is estimated for each day of the year in Table 5.1.

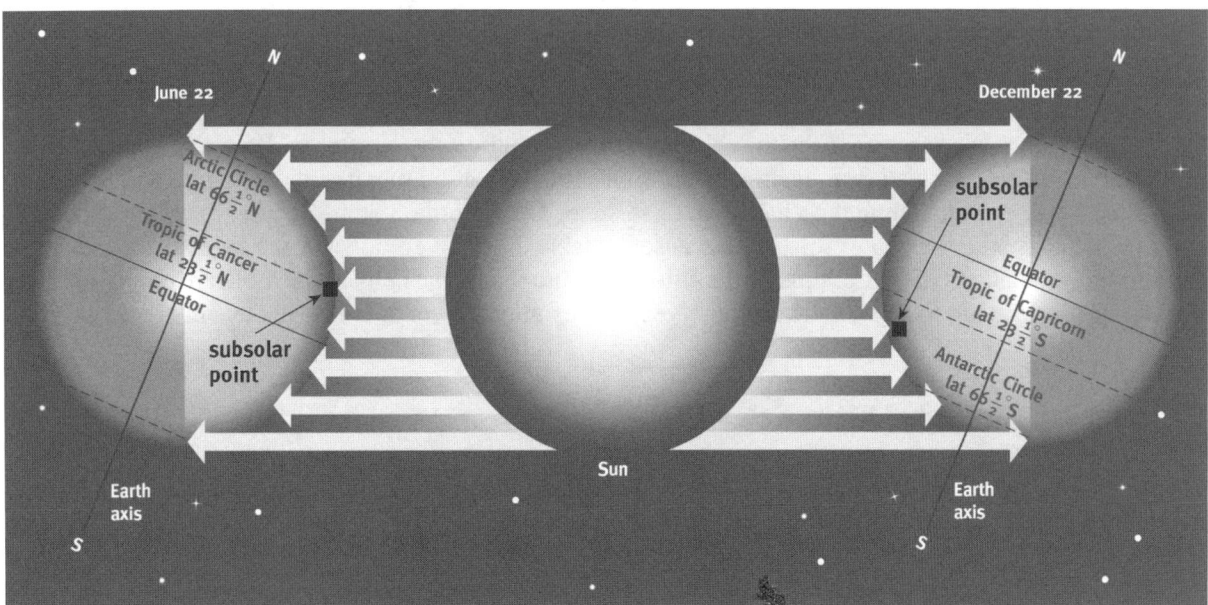

Figure 5.1 Relative positions of Earth and Sun on June 22 and December 22 (Figure 5.2 in H. de Blij, *Physical Geography*, 2005)

Table 5.1 Declination of the Sun Through the Year

Day	JAN	FEB	MAR	APR	MAY	JUN	JUL	AUG	SEP	OCT	NOV	DEC
1	-23°04'	-17°20'	-7°49'	+4°18'	+14°54'	+21°58'	+23°09'	+18°10'	+8°30'	-2°57'	-14°14'	-21°43'
2	-22°59'	-17°03'	-7°26'	+4°42'	+15°12'	+22°06'	+23°05'	+17°55'	+8°09'	-3°20'	-14°34'	-21°52'
3	-22°54'	-16°46'	-7°03'	+5°05'	+15°30'	+22°14'	+23°01'	+17°40'	+7°47'	-3°44'	-14°53'	-22°01'
4	-22°48'	-16°28'	-6°40'	+5°28'	+15°47'	+22°22'	+22°56'	+17°24'	+7°25'	-4°07'	-15°11'	-22°10'
5	-22°42'	-16°10'	-6°17'	+5°51'	+16°05'	+22°29'	+22°51'	+17°08'	+7°03'	-4°30'	-15°30'	-22°18'
6	-22°36'	-15°52'	-5°54'	+6°13'	+16°22'	+22°35'	+22°45'	+16°52'	+6°40'	-4°53'	-15°48'	-22°25'
7	-22°28'	-15°34'	-5°30'	+6°36'	+16°39'	+22°42'	+22°39'	+16°36'	+6°18'	-5°16'	-16°06'	-22°32'
8	-22°21'	-15°15'	-5°07'	+6°59'	+16°55'	+22°47'	+22°33'	+16°19'	+5°56'	-5°39'	-16°24'	-22°39'
9	-22°13'	-14°56'	-4°44'	+7°21'	+17°12'	+22°53'	+22°26'	+16°02'	+5°33'	-6°02'	-16°41'	-22°46'
10	-22°05'	-14°37'	-4°20'	+7°43'	+17°27'	+22°58'	+22°19'	+15°45'	+5°10'	-6°25'	-16°58'	-22°52'
11	-21°56'	-14°18'	-3°57'	+8°07'	+17°43'	+23°02'	+22°11'	+15°27'	+4°48'	-6°48'	-17°15'	-22°57'
12	-21°47'	-13°58'	-3°33'	+8°28'	+17°59'	+23°07'	+22°04'	+15°10'	+4°25'	-7°10'	-17°32'	-23°02'
13	-21°37'	-13°38'	-3°10'	+8°50'	+18°14'	+23°11'	+21°55'	+14°52'	+4°02'	-7°32'	-17°48'	-23°07'
14	-21°27'	-13°18'	-2°46'	+9°11'	+18°29'	+23°14'	+21°46'	+14°33'	+3°39'	-7°55'	-18°04'	-23°11'
15	-21°16'	-12°58'	-2°22'	+9°33'	+18°43'	+23°17'	+21°37'	+14°15'	+3°16'	-8°18'	-18°20'	-23°14'
16	-21°06'	-12°37'	-1°59'	+9°54'	+18°58'	+23°20'	+21°28'	+13°56'	+2°53'	-8°40'	-18°35'	-23°17'
17	-20°54'	-12°16'	-1°35'	+10°16'	+19°11'	+23°22'	+21°18'	+13°37'	+2°30'	-9°02'	-18°50'	-23°20'
18	-20°42'	-11°55'	-1°11'	+10°37'	+19°25'	+23°24'	+21°08'	+13°18'	+2°06'	-9°24'	-19°05'	-23°22'
19	-20°30'	-11°34'	-0°48'	+10°58'	+19°38'	+23°25'	+20°58'	+12°59'	+1°43'	-9°45'	-19°19'	-23°24'
20	-20°18'	-11°13'	-0°24'	+11°19'	+19°51'	+23°26'	+20°47'	+12°39'	+1°20'	-10°07'	-19°33'	-23°25'
21	-20°05'	-10°52'	0°00'	+11°39'	+20°04'	+23°26'	+20°36'	+12°19'	+0°57'	-10°29'	-19°47'	-23°26'
22	-19°52'	-10°30'	+0°24'	+12°00'	+20°16'	+23°26'	+20°24'	+11°59'	+0°33'	-10°50'	-20°00'	-23°26'
23	-19°38'	-10°08'	+0°47'	+12°20'	+20°28'	+23°26'	+20°12'	+11°39'	+0°10'	-11°12'	-20°13'	-23°26'
24	-19°24'	-9°46'	+1°11'	+12°40'	+20°39'	+23°25'	+20°00'	+11°19'	-0°14'	-11°33'	-20°26'	-23°26'
25	-19°10'	-9°24'	+1°35'	+13°00'	+20°50'	+23°24'	+19°47'	+10°58'	-0°37'	-11°54'	-20°38'	-23°25'
26	-18°55'	-9°02'	+1°58'	+13°19'	+21°01'	+23°23'	+19°34'	+10°38'	-1°00'	-12°14'	-20°50'	-23°23'
27	-18°40'	-8°39'	+2°22'	+13°38'	+21°12'	+23°21'	+19°21'	+10°17'	-1°24'	-12°35'	-21°01'	-23°21'
28	-18°25'	-8°17'	+2°45'	+13°58'	+21°22'	+23°19'	+19°08'	+9°56'	-1°47'	-12°55'	-21°12'	-23°19'
29	-18°09'	-8°03'	+3°09'	+14°16'	+21°31'	+23°16'	+18°54'	+9°35'	-2°10'	-13°15'	-21°23'	-23°16'
30	-17°53'		+3°32'	+14°35'	+21°41'	+23°13'	+18°40'	+9°13'	-2°34'	-13°35'	-21°33'	-23°12'
31	-17°37'		+3°55'		+21°50'		+18°25'	+8°52'		-13°55'		-23°08'

Source: http://www.wsanford.com/~wsanford/exo/sundials/DEC_Sun.html

Declination values are negative when the subsolar point is south of the equator and positive when it is north of the equator. A quick examination of the table reveals that the maximum declination value occurs from June 20–3, corresponding with the summer solstice. The minimum value occurs from December 21–4, corresponding with the winter solstice. Because the table does not go down to the level of seconds of arc, it does not pick out the exact time of the solstices. Note that some cells in the table are empty because they do not correspond to days in the calendar year; for example, there are only 30 days in November.

Calculating the Solar Altitude or Elevation

The time of day when the Sun reaches its highest position in the sky is known as **solar noon**. At solar noon the apparent angular height of the Sun above the horizon, the **solar altitude**, is described by the formula:

$$solar\ altitude = 90° - latitude + declination$$

To calculate the solar altitude for a given time and location, the latitude of the site and the solar declination for the day must be known. Let's work through an example for Edmonton, Alberta.

We could determine the solar altitude at solar noon for Edmonton on June 21 by making the following calculations:

1. Determine latitude. Edmonton is at 53° 34′ N.

2. Use Table 1 to determine the declination on June 21st. The declination is +23° 26′.

3. Substitute these values into the formula and employ your knowledge of adding and subtracting degrees and minutes of arc from Unit 3. (Hint: express 90° as 89° 60' so that you have both degrees and minutes to subtract 53° 34′ from).

$$\text{solar altitude} = 90° - \text{latitude} + \text{declination}$$

$$\text{solar altitude} = 89° 60' - 53° 34' + 23° 26'$$

$$\text{solar altitude} = 59° 52'$$

On June 21, in Edmonton, Alberta at solar noon, the Sun will be 59°52′ above the horizon (Figure 5.2). Note that since June 21 is the summer solstice, this is the highest the Sun will rise above the horizon at this location. Edmonton does not ever experience perpendicular solar radiation.

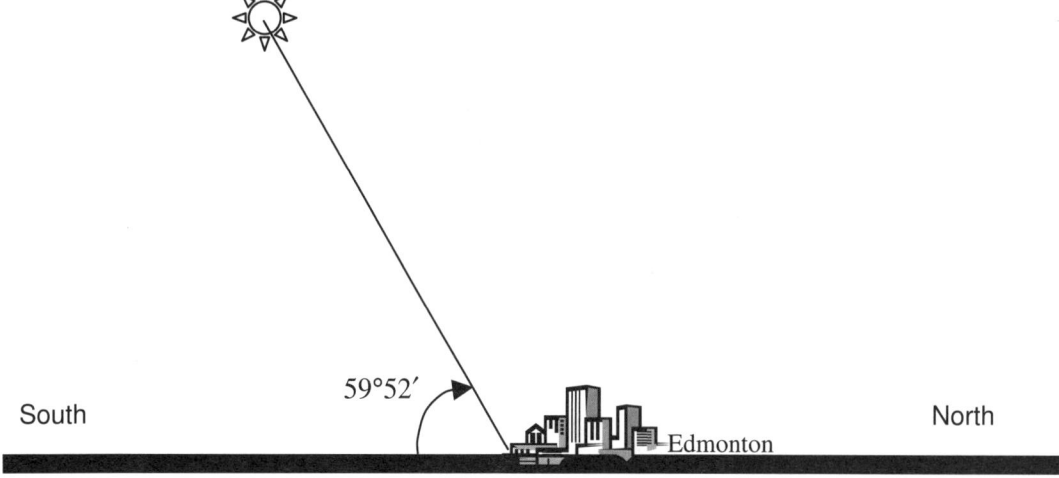

Figure 5.2 The solar altitude at solar noon in Edmonton, Alberta on June 21st is 59°52′. The solar altitude is the angle from the horizon to the noon sun.

Try Your Hand at Some Questions

1. Calculate the solar altitude for Yellowknife, Northwest Territories (Latitude 62° 28′ N) on the winter solstice (December 21st).

2. Calculate the solar altitude for Halifax, Nova Scotia (Latitude 44° 38′ N) on the winter solstice.

1. Yellowknife's solar altitude on Dec. 21st = 4°06'.
2. Halifax's solar altitude on Dec. 21st = 21°56'.

Radiation and the Heat Balance of the Atmosphere: A Supplement to Unit 7

The purpose of this section is to provide supplementary background information to help you understand what happens as solar and terrestrial radiation are emitted and interact with the atmosphere and ground surface of our planet.

Information, sample calculations, and practice questions are provided on the:

1. emission and receipt of radiant energy, including:
   a. electromagnetic radiation
   b. the absolute temperature scale
   c. Wien's Displacement Law
   d. the Stefan-Boltzmann Law
   e. total energy emission from a blackbody
   f. the Inverse Square Law
   g. the Cosine Law of Illumination

2. radiation and energy balances, including:
   a. components of the radiation balance
   b. reflectivity and transmission
   c. interpretation of radiation balance data over annual and diurnal timeframes
   d. components of the energy balance
   e. interpretation of energy balance data over annual and diurnal timeframes
   f. atmospheric greenhouse effect

Electromagnetic Radiation

Your body is constantly both emitting and being bombarded by electromagnetic radiation (radiant energy, or radiation). If you turn a radio on it will broadcast music because it is receiving specific wavelengths of radiant energy from the station transmitter. The light you see coming from the sun is radiation. This booklet is emitting radiant energy and so are your desk, your shoes, and your highlighter pen. In fact, all objects above absolute zero (-273°C, the temperature at which all molecular motion ceases) emit electromagnetic energy.

Canadians use Système Internationale (metric) units of measurement, and electromagnetic radiation—like other forms of energy—is measured in Joules (J). Sometimes we will be interested in the *flux* of energy, or the rate of energy flow per unit of time. This is given in Joules per second (J/s, or J s^{-1}) which are also known as Watts (W). Other times, we will be interested in the *flux density*. Like population density, when a certain number of people live within a square kilometre of land, flux density refers to a rate of energy flow per unit area. It is cited in Watts per square metre (W/m^2 or W m^{-2}).

Electromagnetic radiation (EMR) has both wave and particle properties. While some scientific studies consider particle (photon) characteristics, in the field of climatology it is useful to examine EMR's wave properties. Electromagnetic waves have no mass, but they can transmit energy from place to place. In the vacuum of space they travel at the speed of light (3 x 10^8 m/s^{-1}, or 300 million metres per second). These waves are commonly described in terms of their wavelength and frequency.

The wavelength is symbolized by the Greek letter *lambda* (λ) and measured as the distance between successive crests or troughs. Very small wavelengths are measured in micrometres (μm – with the Greek letter *mu* – μ). These represent a distance of one-millionth of a metre (10^{-6} m). Larger wavelengths may be measured in metres.

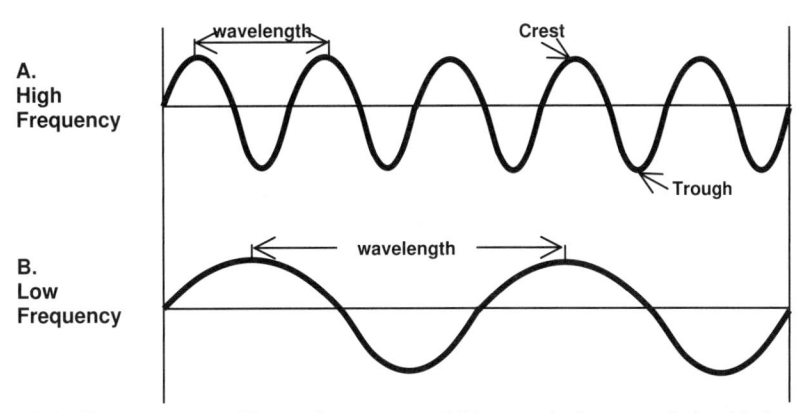

Figure 7.1. Two waves travelling at the same speed illustrate the inverse relationship between wavelength and frequency. Wave A has a small wavelength; many crests will pass a given point each second, resulting in a high frequency. Fewer crests will pass a given point each second in the longer wavelength of Wave B, resulting in a lower frequency.

The frequency refers to the number of wave crests that pass a given point per second. Frequency is inversely related to the wavelength. Both of the waves illustrated in Figure 7.1 travel at the same speed (the speed of light). The wave with the shorter wavelength will have more crests pass a given point each second than the wave with the longer wavelength, therefore, it will have a higher frequency. Conversely, longer wavelengths produce lower frequencies.

Electromagnetic radiation varies along a continuum of extremely small to very large wavelengths known as the *electromagnetic spectrum*. The spectrum is divided into categories or spectral bands, as illustrated in Figure 7.2. The human eye registers only a small portion of the spectrum called *visible light*, which ranges from 0.4–0.7μm. Our eyes don't sense shorter ultraviolet wavelengths or longer infrared or radio bands. All objects above absolute zero radiate a range of wavelengths. The Sun, for instance, radiates wavelengths from the ultraviolet through to thermal infrared bands. The Earth radiates wavelengths from near infrared to the thermal infrared parts of the spectrum. So why do we say that the Sun emits shortwave radiation while the Earth emits longwave energy?

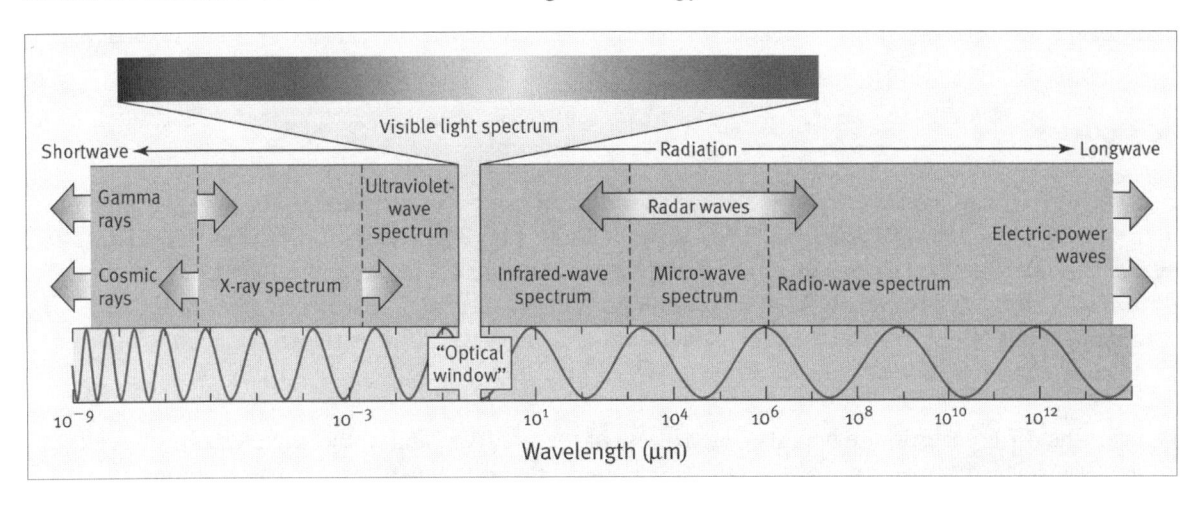

Figure 7.2. The complete (electromagnetic) spectrum

Careful observation and measurement of the energy emitted by objects reveals that their surface temperature determines both the wavelengths and intensity of the electromagnetic radiation they emit. To explore this in more detail, we first need to learn about the absolute temperature scale.

33

The Absolute or Kelvin Temperature Scale

Any object with a temperature above absolute zero emits radiant energy. No energy is radiated at absolute zero because this is the temperature at which all molecular motion stops. Absolute temperature (T) is measured in kelvins (K). No degree symbols are used; the scale is based on *absolute* zero.

The size of a Celsius degree is the same as a kelvin, so these temperature scales are similar except for their starting points. Absolute zero (0K) corresponds to -273.15°C, although we'll round that to -273°C. To convert temperatures from °C to K simply add 273. To convert temperature from K to °C simply subtract 273.

$$K = °C + 273$$

$$°C = K - 273$$

The Sun's surface temperature has been estimated to be 5770 K. What is that in °C? The formula is °C = K – 273. Substituting 5770K into the formula gives °C = 5770K – 273 = 5497°C.

Web activity

If you are curious about how we estimate the surface temperature of the Sun, keep reading the information below on Plank curves and Wien's law, then check out the Cornell University Astronomy site at http://curious.astro.cornell.edu/question.php?number=126.

A Question for You

It's your turn! Earth's average surface temperature has been estimated to be 15°C. What is that temperature when expressed in Kelvin? The answer is given below, but no cheating! Work out your response and then have a look.

The formula is K = °C + 273. Substituting in 15°C gives
K = 15°C + 273 = 288K.

Planck curve

Any body at a given surface temperature radiates energy in not just one wavelength, but over a range of wavelengths as shown below by the Plank curves for the Sun (Fig. 7.3) and the Earth (Fig. 7.4). These graphs illustrate the intensity of energy emitted each second from each square metre of a body's surface (the *flux density*) for given wavelengths. The peak of each curve indicates at which wavelength the maximum emissions occur. For the Sun, the peak emission is at 0.5μm (the arrow in Fig. 7.3), while for Earth, the peak emission is closer to 10μm (the arrow in Fig. 7.4).

What causes the differences in the heights (intensities) of these curves and in the wavelengths of peak emissions? The surface temperature of the radiating bodies creates the differences. The relationship between temperature and the wavelength of peak emission (λ_{max}) has been described in Wien's Law. The relationship between temperature and the intensity of emitted radiation has been described by the Stefan-Boltzmann Law. These two important laws form the basis for understanding the role of radiant energy in climatology.

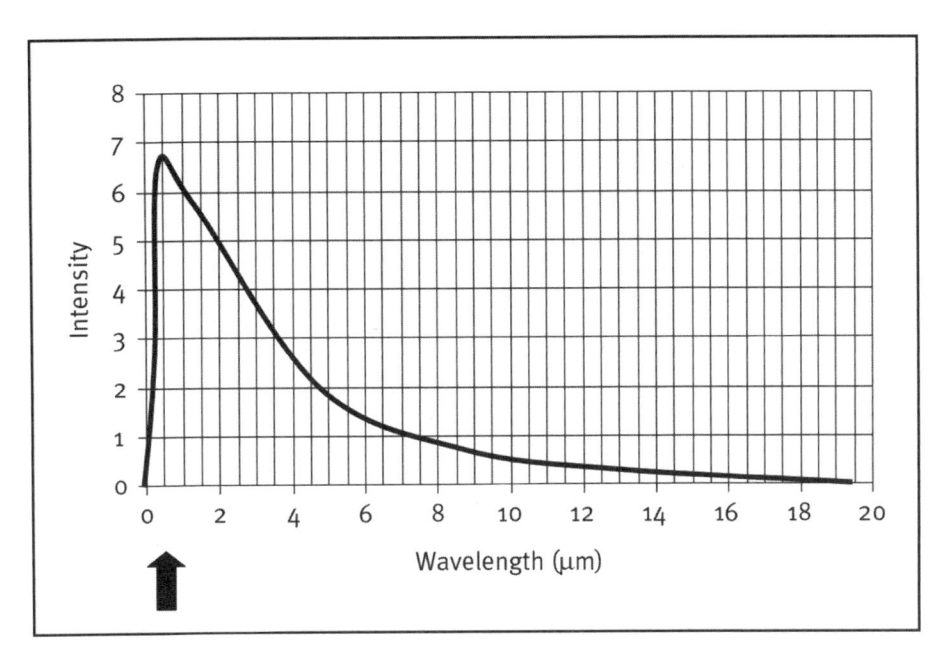

Figure 7.3. The intensity of radiation emitted by the sun in different wavelengths. The Sun's surface temperature is 5770 K.

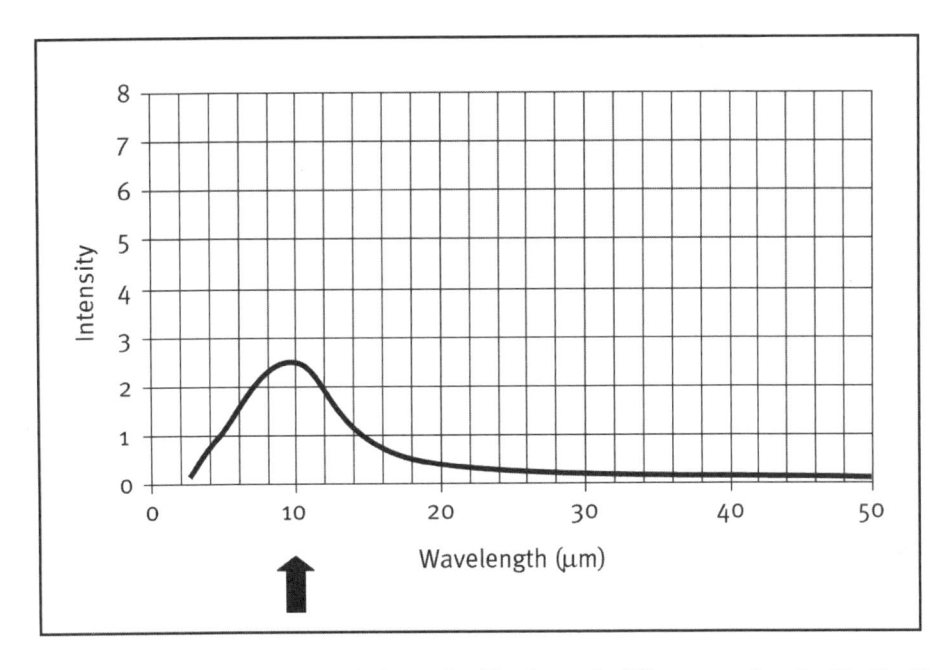

Figure 7.4. The intensity of radiation emitted by the sun in different wavelengths. The Earth's surface temperature is 300 K.

Wien's Law

Objects radiate energy over a range of wavelengths. Wien's Law identifies the wavelength at which most energy is emitted (λ_{max}) by a given body at a given surface temperature. Figure 7.4 above presents the wavelength emissions of Earth with a surface temperature of 300K. As the figure illustrates, this body radiates wavelengths that range from 2.5µm to > 50µm, but most of its energy is radiated at a wavelength of roughly above 10μm. This is the peak energy emission (λ_{max}) for a surface temperature of 300 K. That is what Wien's Law identifies.

Wien's Law can be stated as the following formula:

$$\lambda_{max} = \frac{2898 \mu m K}{T}$$

where T is the absolute temperature of the radiating body and λ_{max} is in μm.

Using this equation we can calculate the peak energy emission (λ_{max}) for any surface if we know its temperature. For instance, to determine the precise λ_{max} for the body depicted in Figure 7.4, we would substitute its absolute temperature into the equation.

$$\lambda_{max} = \frac{2898 \mu m K}{300 K} = 9.66 \mu m$$

Therefore, the wavelength of peak energy emission for Earth is 9.66μm.

Question for Practice

The Sun has a surface temperature of 5770 K. At what wavelength is the sun emitting most of its radiant energy? You may verify your work with the answer at the end of this unit.

Examine Figure 7.5 showing the influence of temperature on the peak energy emission of bodies. What pattern do you see portrayed in this graph? As the temperature of the radiating body increases, what happens to its λ_{max}?

Is this a direct relationship or an inverse relationship?

As the temperature of the radiating body increases, the peak wavelength decreases. This, therefore, is an *inverse* relationship. As one variable increases the other decreases.

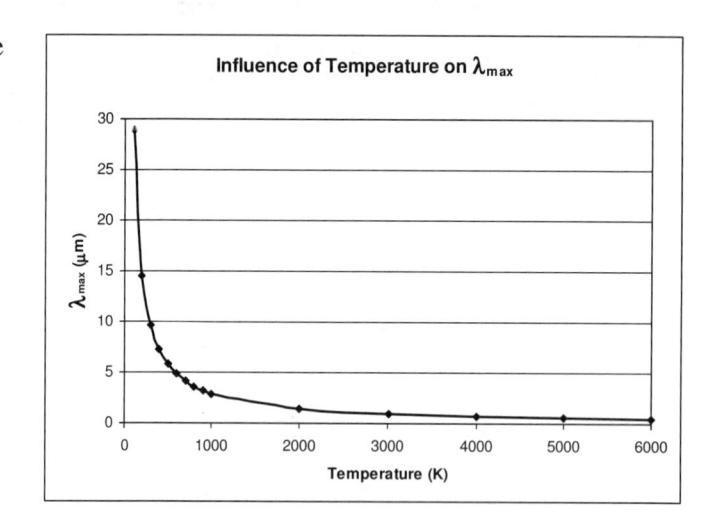

Figure 7.5. Graphical depiction of Wien's Displacement Law.

Stefan-Boltzmann Law

Blackbodies (perfect emitters/absorbers of radiant energy) emit radiant energy in direct proportion to the 4th power of their surface temperature. From the Planck curve, the total amount of radiant energy emitted in all wavelengths at a given temperature is given by the area under the curve (See Figures 7.3 and 7.4).

This law is expressed mathematically as the Stefan-Boltzmann equation:

$$\textit{Intensity of Energy Emitted, } E = \sigma T^4$$

where σ (lower case Greek letter *sigma*) is the Stefan-Boltzmann constant (5.67 x 10^{-8} $Wm^{-2}K^{-4}$), T is expressed in Kelvin, and the flux density is expressed in Wm^{-2}.

In a first-year course we will assume that all objects are blackbodies, although it is a simplification of reality. This assumption will be corrected in second-year classes where an emissivity term will be added to the equation.

We can use the Stefan-Boltzmann law to calculate the radiative flux density of a blackbody if we know that body's surface temperature. For example, to calculate the radiative flux density of the Sun, we assume that the Sun is a blackbody and has a surface temperature of 5770 K. Then we plug this temperature value into the equation and solve. Students who have a lot of science in their background may make shortcuts with the following calculations. It may be worthwhile to follow all the steps the first time and make the shortcuts in later exercises.

Step 1   Energy Intensity = σT^4

Step 2   Energy Intensity = $\sigma (5770\ K)^4$

Step 3   Energy Intensity = $\sigma\ 5770^4\ K^4$

Step 4   Energy Intensity = $\sigma (5.77 \times 10^3)^4\ K^4$

We will put the 'digit' part of the number to the 4th power, then the 'exponent' part of the number to the 4th power. Remember that 5.77 must be multiplied by itself four times (5.77 to the 4th power) = 1108.4. Then we must put 10^3 to the 4th power (multiple 10^3 by itself four times; the shortcut is to multiply the exponents) = 10^{12}.

Step 5   Energy Intensity = $\sigma (1108.4 \times 10^{12}\ K^4)$

By convention, we write numbers in scientific notation with the decimal after the first digit. So let's convert this number.

Step 6   Energy Intensity = $\sigma (1.1084 \times 10^{15}\ K^4)$

Now we substitute the value of σ into the equation. (We didn't need to spend the time writing it out in full until now.)

Step 7   Energy Intensity = (5.67 x 10^{-8} $Wm^{-2}K^{-4}$) (1.1084 x 10^{15} K^4)

You can see that the Kelvin units cancel out (K^{-4} and K^4).

Step 8   Energy Intensity = (5.67 x 10^{-8} Wm^{-2}) (1.1084 x 10^{15})

Now we multiply the 'digit' parts of the numbers together and add the 'exponent' parts.

Step 9   Energy Intensity = (5.67)(1.1084) x 10^{-8+15} Wm^{-2}

Step 10   Energy Intensity = (5.67)(1.1084) x 10⁷ Wm⁻²

Step 11   Energy Intensity = 6.2846 x 10⁷ Wm⁻²

      Now we'll check for significant figures. What is the least number of sig. figs. in our original data? Three? Notice that we don't round off until the final answer. If we rounded at every step we would lose considerable accuracy.

Step 12   Energy Intensity = 6.29 x 10⁷ Wm⁻²

So the Sun emits radiation with an intensity of 6.29 x 10⁷ Wm⁻².

Figure 7.6 depicts how the intensity of emitted energy from a blackbody increases as its absolute temperature increases. Note that the values on the vertical axis increase by a factor of 10 at each increment. This is referred to as a *logarithmic* scale.

What pattern do you see portrayed in this graph? As the temperature of the radiating body increases, what happens to the intensity of energy emission? Is this a direct relationship or an inverse relationship? As the temperature increases, so does the flux density, therefore this is a direct relationship.

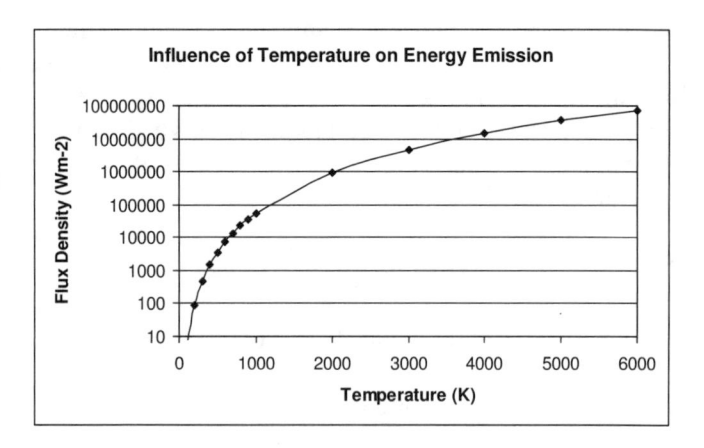

Figure 7.6. Graphical depiction of the Stefan-Boltzmann Law.

Calculating Total Energy Emission from a Blackbody

The total amount of energy an object radiates is determined by multiplying its surface area by the average intensity of energy emission of the object.

A surface area measured in m² multiplied by an emission intensity expressed in Wm⁻² will result in a total energy emission given in W.

The Sun has a surface area of 6.09 x 10¹⁸ m². In the preceding section we calculated the average intensity of energy emission for the Sun as 6.29 x 10⁷ Wm⁻². Based on these values, it follows that:

Step 1   Total Energy Emission = (Surface Area) (Intensity of Energy Emission)

Step 2   Total Energy Emission = (6.09 x 10¹⁸ m²) (6.29 x 10⁷ Wm⁻²)

      You can see that the m² terms will cancel (m² and m⁻²). We multiply the 'digit' parts of the numbers and then add the 'exponent' parts.

Step 3   Total Energy Emission = (6.09) (6.29) x 10¹⁸⁺⁷ W

Step 4   Total Energy Emission = 38.3061 x 10²⁵ W

      Rounding to three sig. figs. and moving the decimal place gives us...

Total Energy Emission = 3.83 x 10²⁶ W

So the total energy emission of the Sun is 3.83 x 10²⁶ W.

Let's recap.

- We applied the surface temperature of the Sun to the Stefan-Boltzmann Law to calculate the intensity of energy given off by the Sun (the amount of energy it emits every second from every square metre of its surface).

- We multiplied that figure by the Sun's surface area to determine its total energy output per second.

Now we want to investigate how much of the Sun's energy output is intercepted by Earth. Think of the Sun as a point source radiating energy in all directions, like a light bulb. If you are reading a newspaper and your only source of light is one bulb, how close you stand to that bulb influences your reading light. In the same way, how close Earth is to the Sun determines the intensity of solar energy we intercept. There is a *distance–decay factor* that is described by the Inverse Square Law.

Calculating the Solar Constant with the Inverse Square Law

This law describes the energy relationships between any radiating and intercepting bodies. In our examples the radiating body will be the Sun and the receiving body will be the Earth, but the law describes the relationship between any two bodies. According to the Inverse Square Law, the intensity of radiation received at distance r from a body with a total energy emission of Q is given by:

$$\text{Receipt Intensity (Flux Density)} = \frac{Q}{4\pi r^2}$$

If Q is measured in Watts and r is measured in metres, the receipt intensity will have dimensions of W m⁻². The calculated receipt intensity is the intensity of radiative energy *intercepted by a plane at right angles to the incoming radiation*.

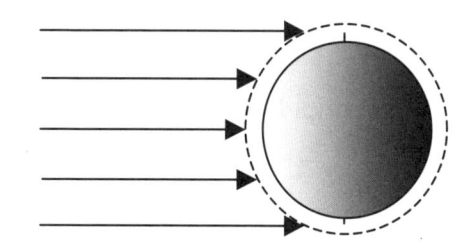

This is known as the **solar constant**.

Enlarging a view of the outer atmosphere perpendicular to the solar beam may help visualize the previous description of the solar constant. The solar constant refers to the amount of energy flowing each second across a 1 m² plane at the outer edge of the atmosphere perpendicular to the solar beam.

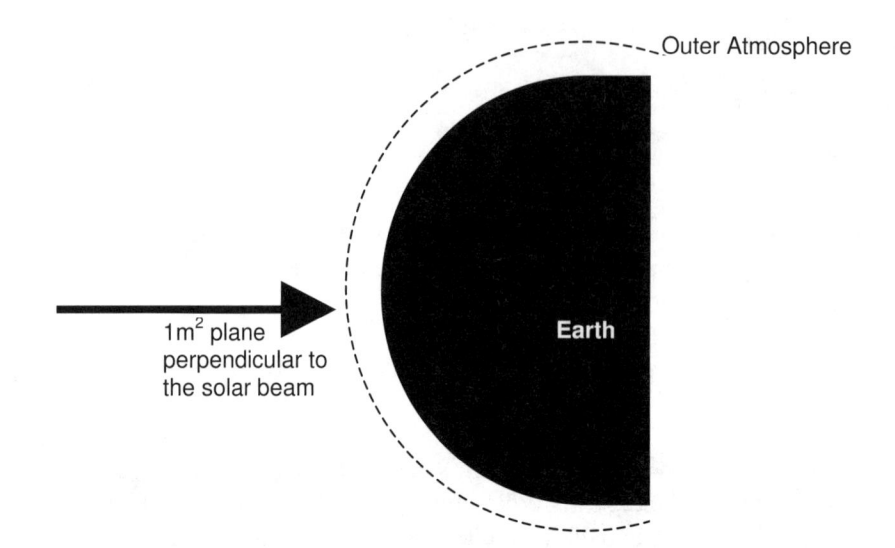

To calculate the intensity of solar radiation intercepted by the Earth when it is at its average distance from the Sun, assume that the total energy emission of the Sun (Q) is 3.83 x 10^{26} W, as calculated in the preceding section. The average Sun–Earth distance (r) is 1.497 x 10^{11}m, hence:

Step 1 $\quad$ Receipt Intensity $= \dfrac{Q}{4\pi r^2}$

Step 2 $\quad$ Receipt Intensity $= \dfrac{3.83 x 10^{26} W}{4\pi (1.497 x 10^{11} m)^2}$

Step 3 $\quad$ Receipt Intensity $= \dfrac{3.83 x 10^{26} W}{4\pi (1.497^2 x 10^{22} m^2)}$

Step 4 $\quad$ Receipt Intensity $= \dfrac{3.83 x 10^{26} W}{4\pi (2.241 x 10^{22} m^2)}$

Step 5 $\quad$ Receipt Intensity $= \dfrac{3.83 x 10^{26} W}{4(3.1416)(2.241 x 10^{22} m^2)}$

Step 6 $\quad$ Receipt Intensity $= \dfrac{3.83 x 10^{26} W}{(12.6)(2.241 x 10^{22} m^2)}$

Step 7 $\quad$ Receipt Intensity $= \dfrac{3.83 x 10^{26} W}{28.2 x 10^{22} m^2}$

Step 8 $\quad$ Receipt Intensity $= \dfrac{3.83 x 10^{26} W}{2.82 x 10^{23} m^2}$

Step 9   Receipt Intensity $= \dfrac{3.83}{2.82} \times 10^{26-23} Wm^{-2}$

Step 10   Receipt Intensity $= 1.36 \times 10^3 Wm^{-2}$

Our calculations indicate that the average intensity of solar radiation received at the outer edge of Earth's atmosphere perpendicular to the solar beam is 1360 Wm^{-2}. Satellite measurements of this solar constant put the value at 1372 Wm^{-2}.

> **Question for Practice**
>
> Given that the average Sun–Jupiter distance is 7.786 x 10^{11} m, calculate the intensity of solar radiation that is intercepted by Jupiter. The answer is given at the end of the unit.

Advanced Topic: Calculating the Extraterrestrial Radiation with the Cosine Law of Illumination

Not all solar radiation is intercepted at right angles at the top of the atmosphere. Subsolar points within the tropics are momentarily able to intercept solar radiation at right angles when the noon sun angle is overhead. For all times other than noon and for all extra-tropical locations, solar radiation is intercepted at acute angles at the top of the atmosphere and is termed extra-terrestrial radiation. The Cosine Law of Illumination describes how the intensity of extraterrestrial radiation is reduced according to the angle at which it is intercepted:

$$K_{ex} = I_o \cos Z$$

Or we can choose the alternative formula that yields the same result:

$$K_{ex} = I_o \sin A$$

where K_{ex} is extraterrestrial radiation in Wm^{-2}, I_o is the solar constant in Wm^{-2}, Z is the zenith angle in degrees, and A is the solar altitude in degrees.

When the sun is overhead, Z = 0° and cos 0° = 1; A = 90° and sin 90° = 1. This means that extra-terrestrial radiation will equal the solar constant. On the other hand if it is sunrise or sunset, Z = 90° and cos 90° = 0; A = 0° and sin 0° = 0. This means that extraterrestrial radiation will not be available at these times.

Using the second formula (which follows from the solar altitude calculations we made in Unit 5), for example, the noon sun angle for Vancouver (49° N) on the June solstice can be calculated as follows:

K_{ex} = I_o sin A
     = 1360 Wm^{-2} x sin 64.5°
     = 1360 Wm^{-2} x 0.902585284
     = 1227.5 Wm^{-2}

> **Question for Exercise**
>
> The noon sun angle for Vancouver on the December solstice is given as 17.5°. What intensity of solar radiation is intercepted at the top of the atmosphere? The answer is shown at the end of this unit.

Surface Radiation Balance

When extraterrestrial radiation interacts with the atmosphere, it will partly be reflected, scattered or absorbed by the atmosphere. A portion of it is transmitted to Earth's surface, and may arrive as direct solar radiation or diffuse radiation. This proportion, termed transmissivity, is calculated as:

$$\tau = \frac{K\downarrow}{K_{ex}}$$

where τ is transmissivity and is unit-less; and $K\downarrow$ is incoming solar radiation at Earth's surface in Wm^{-2}.

A portion of incoming solar radiation is reflected back depending on the physical properties of the surface. The proportion of reflected solar radiation to incoming solar radiation is termed surface reflectivity or albedo. This can be calculated as follows:

$$\alpha = \frac{K\uparrow}{K\downarrow}$$

where α is surface reflectivity or albedo and $K\uparrow$ is reflected solar radiation in Wm^{-2}.

This relationship may also be expressed as:

$$K\uparrow = \alpha\, K\downarrow$$

The albedo of different surfaces on Earth varies greatly, ranging from about 3 per cent over forests to about 95 per cent over fresh snow. Due to this property some surfaces do not gain as much heat from the Sun as they should because a lot of incoming energy is being reflected back. For instance, if you have incoming solar radiation equal to 1145 Wm^{-2} on a fresh snowfield, we can determine how much of it is reflected away from the surface. Here is the calculation:

$\alpha = \dfrac{K\uparrow}{K\downarrow}$, since the albedo for fresh snow is known to be 95 per cent, this equation becomes:

$\dfrac{95}{100} = \dfrac{K\uparrow}{1145\, Wm^{-2}}$, and this leads to $\quad 1145\, Wm^{-2}\left(\dfrac{95}{100}\right) = K\uparrow$

So, $K\uparrow = 1087.75\, Wm^{-2}$. We see that very little energy will remain to heat the snowfield and it will likely not thaw quickly due to this reason.

Exercise

Try determining the amount of energy reflected off a sandy surface with an albedo of 36 per cent and incoming solar radiation of 1237 Wm^{-2}. The answer is at the end of the unit.

Also arriving at the surface is **incoming longwave radiation** emitted by the atmosphere. **Outgoing longwave radiation** is emitted by the Earth's surface; this escapes into the atmosphere and space. Depending on how large each of all the radiative components is, some amount of energy called net radiation is gained or lost at the surface. Overall, the surface radiation balance is given by:

$$Q^* = K\downarrow - K\uparrow + L\downarrow - L\uparrow$$

where Q* is net radiation in Wm^{-2}, L$\downarrow$ is incoming longwave radiation in Wm^{-2}, and L$\uparrow$ is outgoing longwave radiation in Wm^{-2}.

Note the surface radiation balance is made up of shortwave radiation and longwave radiation. The surface balance specific to solar radiation is given as:

$$K^* = K\downarrow - K\uparrow$$

where K* is net solar radiation in Wm^{-2}.

Similarly the surface balance specific to longwave radiation is:

$$L^* = L\downarrow - L\uparrow$$

where L* is net longwave radiation in Wm^{-2}.

Net radiation may also be obtained from the following equation:

$$Q^* = K^* + L^*.$$

We will work through a hypothetical example to help explain the surface radiation balance. In the table below, the sky is cloudless and the components of the radiation balance were measured except for the two missing values. All the values are in Wm^{-2}.

Time	1200	1400
K$\downarrow$	991	1008
K$\uparrow$	?	201
L$\downarrow$	411	420
L$\uparrow$	428	?
Q*	776	794

We wish to know how much of the energy was reflected away from the Earth's surface at noon. We do not know the albedo but fortunately we have all the other components in the surface radiation balance.

Here is how we can solve this:

Write the surface radiation balance equation and then substitute all the values that are available except for the missing variable.

$$Q^* = K\downarrow - K\uparrow + L\downarrow - L\uparrow$$

$$776 \text{ Wm}^{-2} = 991 \text{ Wm}^{-2} - K\uparrow + 411 \text{ Wm}^{-2} - 428 \text{ Wm}^{-2}$$

We then isolate the unknown variable.

$$K\uparrow = 991 \text{ Wm}^{-2} + 411 \text{ Wm}^{-2} - 428 \text{ Wm}^{-2} - 776 \text{ Wm}^{-2} = 198 \text{ Wm}^{-2}$$

See if you can arrive at the value of 433 Wm^{-2} for outgoing longwave radiation at 1400 hrs.

Surface Energy Balance

Net radiation accumulated at the surface of Earth (Q^*) is dissipated by non-radiative means, i.e. **convection**, **conduction**, **evaporation**, or **photosynthesis**. What proportion is dissipated by each of these processes depends on the nature of the surface. For example a wet surface may use most of the energy it receives to evaporate water. In contrast, the energy received by a dry desert surface with no water to evaporate will primarily be dissipated by the convective heating of the air above the surface. How a surface dissipates the net radiative energy it receives is summarized by the surface's energy balance. For many terrestrial surfaces this can be stated as:

$$Q^* = Q_H + Q_E + Q_G$$

Where Q_H is sensible heat in Wm^{-2}, the energy used to convectively heat the air above the surface, Q_E is latent heat in Wm^{-2}, the energy used to evaporate water from the surface, and Q_G is ground heat in Wm^{-2}, the energy used to conductively heat the soil below the surface.

However, ground heat on all surface types is usually small because Earth's conductivity is poor. Annually, ground heat conducted downward or upward amounts to zero storage.

What happens to radiant energy as it passes through the atmosphere?

Most of the variable gases in the atmosphere selectively absorb longwave radiation and re-emit it again (see Figure 7.7 below). While these gases are largely transparent to solar radiation, they delay the release of the longwave radiation from the atmosphere to space by re-emitting a portion of the longwave radiation back to Earth. This delay is called the **atmospheric greenhouse effect**. This is a natural phenomenon that helps to keep the atmosphere warm in the absence of the sun at night. Because these gases contribute to the atmospheric greenhouse effect, they are called greenhouse gases (GHGs). The strongest absorbers of infrared radiation are water and carbon dioxide. As Figure 7.7 indicates, nitrous oxide and methane are also greenhouse gases. If the amounts of GHGs are increased by human activity, many scientists are concerned that they can disrupt the radiation and energy balances and possibly lead to global warming.

None of the atmospheric gases can strongly absorb longwave radiation between 8 and 11 μm. This interval in the electromagnetic spectrum, which allows longwave radiation to escape into space, is called the **atmospheric window**. However, cloud cover is capable of effectively blocking this atmospheric window thereby maintaining warmer surface temperatures.

Note that oxygen and **stratospheric ozone** absorb most of the harmful UV radiation in the upper atmosphere, which is a good thing for living organisms on Earth. Stratospheric ozone is generated naturally by a sequence of chemical reactions involving atomic oxygen (O), molecular oxygen (O_2), and ozone (O_3). On the contrary, **tropospheric ozone** absorbs longwave radiation in the lower atmosphere and contributes to warming. The accumulation of tropospheric ozone comes about when in the presence of sunlight, oxygen (O_2) reacts with pollutants; namely nitrogen dioxide (NO_2) and hydrocarbons.

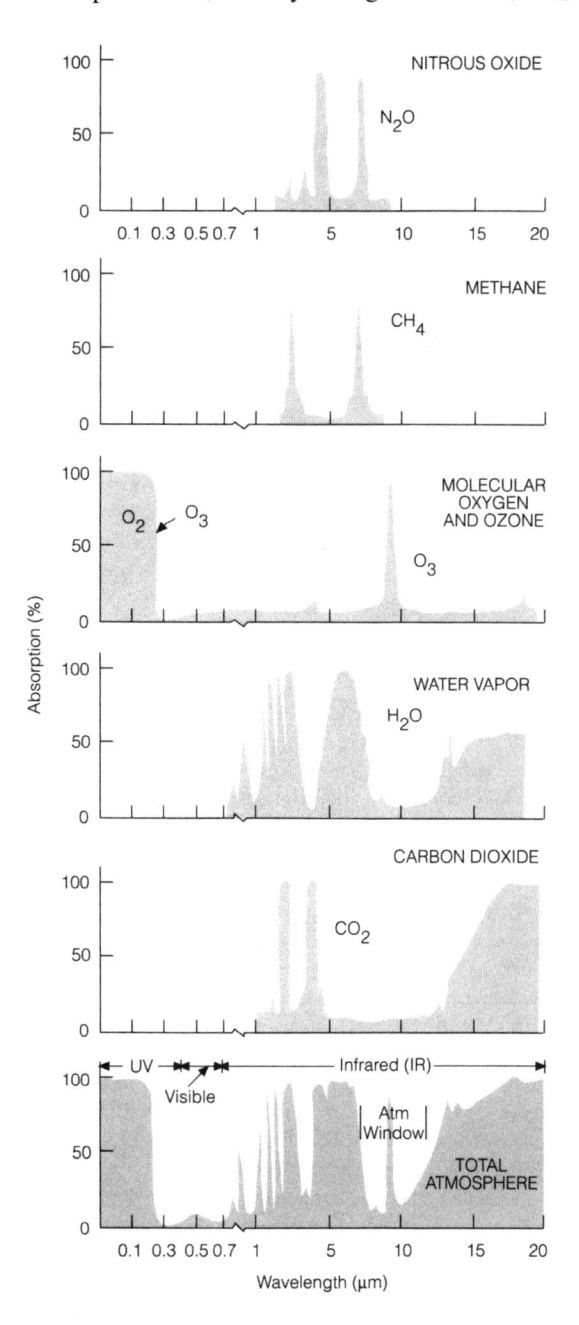

Figure 7.7 Absorption of radiation by atmospheric gases across the electromagnetic spectrum. From *Meteorology Today (with 1pass for MeteorologyNOW(T))* 8th edition by Ahrens. © 2007. Reprinted with permission of Brooks/Cole, a division of Thomson Learning: www.thomsonrights.com. Fax 800 730-2215.

45

Solutions for Practice Questions

Sun's wavelength for peak radiant energy emission

$$\lambda_{max} = \frac{2898 \mu m K}{5770 K} = 0.50 \mu m$$

Jupiter's intercepted solar radiation

$$\text{Receipt Intensity} = \frac{3.83 \times 10^{26} W}{4\pi (7.786 \times 10^{11} m)^2} = 5.03 \times 10^1 Wm^{-2}$$

Extraterrestrial radiation for Vancouver on December solstice

$$\begin{aligned} K_{ex} &= I_o \sin A \\ &= 1360 \, Wm^{-2} \times \sin 17.5° \\ &= 1360 \, Wm^{-2} \times 0.3007058 = 409.0 \, Wm^{-2} \end{aligned}$$

Reflected energy from a sandy surface

$$K\uparrow = 1237 \, Wm^{-2} \left(\frac{36}{100}\right) = 445.32 \, Wm^{-2}$$

Atmospheric and Surface Temperature: A Supplement to Unit 8

The purpose of this section is to help you see how temperature is derived for different time intervals and also to show how such values can be used to produce a temperature map.

Information, sample calculations, and practice questions are provided on:

1. temperature calculations:
    a. mean daily temperature
    b. mean annual temperature range
    c. mean monthly temperature
    d. mean annual temperature
2. temperature patterns including:
    a. temperature maps
    b. isotherms
    c. temperature gradients

Temperature

Temperature is a measure of the kinetic energy contained in a substance. This energy causes molecules to vibrate in a random fashion. The higher the kinetic energy contained in a substance, the higher its temperature reading will be. Changes in temperature result from a **flux convergence** or **divergence** of sensible heat through an object. A flux convergence occurs when sensible heat entering a substance exceeds that which is leaving. This results in warming. Conversely, a flux divergence occurs when the sensible heat leaving a substance is greater than that which is entering. A cooling effect is produced by this situation. In the climate system, convection is responsible for air temperature changes while conduction is associated with soil temperature changes.

A min–max thermometer can measure the respective minimum and maximum temperatures of a given day. From such daily measurements, temperature values for other time scales can be derived.

The **mean daily temperature** is the mean of the maximum and minimum temperature for a given day:

$$\overline{T}_{daily} = \frac{T_{max} + T_{min}}{2}$$

where $\overline{T}_{daily}$ is the mean daily temperature in degrees Celsius, T_{max} is the minimum daily temperature in degrees Celsius, and T_{min} is the maximum daily temperature in degrees Celsius.

In turn, the **mean monthly temperature** is calculated as the mean of all mean daily temperatures for a given month:

$$\overline{T}_{monthly} = \frac{\Sigma \overline{T}_{daily}}{N}$$

where $\overline{T}_{monthly}$ is the mean monthly temperature in degrees Celsius for a given month, $\Sigma \overline{T}_{daily}$ is the sum of all mean daily temperatures in degrees Celsius for a given month, and N is the total number of days in a given month.

Finally, the **mean annual temperature** is derived from the mean of all mean monthly temperatures in a given year:

$$\overline{T}_{annual} = \frac{\Sigma \overline{T}_{monthly}}{12}$$

where $\overline{T}_{annual}$ is the mean annual temperature in degrees Celsius for a given year, $\Sigma \overline{T}_{monthly}$ is the sum of all mean monthly temperatures in degrees Celsius for a given year, and 12 is the number of months in any given year.

Although two locations can have similar mean temperature values, they may sometimes show a completely different picture in terms of temperature ranges. This may largely result from the nature of the surface over which the temperature is being read. For example, coastal areas tend to have high mean temperatures but have relatively lower temperature ranges. Interior locations have low mean annual temperatures in high latitudes and high mean annual temperatures in low latitudes. The **annual temperature range** is calculated as follows:

$$\Delta T = T_{annual\ max} - T_{annual\ min}$$

where ΔT is the annual temperature range, $T_{annual\ max}$ is the highest mean monthly temperature value in a given year, $T_{annual\ min}$ is the lowest mean monthly temperature value in the given year.

Example

On a given day a maximum temperature of 22°C occurs in the afternoon and a minimum temperature of 2°C is recorded after sunrise. What is the mean daily temperature?

Answer

$$\overline{T}_{daily} = \frac{T_{max} + T_{min}}{2} = \frac{22\ °C + 2\ °C}{2} = \frac{24\ °C}{2} = 12\ °C$$

> **Try This**
>
> A winter day has a maximum temperature of 12°C and a minimum temperature of -2°C. What is the mean daily temperature? Work out your response and then have a look. The answer is at the end of this unit.

Another example

For one year, the mean monthly temperatures in degrees for January to December are: 2, 1, 2, 5, 13, 20, 22, 26, 21, 17, 11, and 4. Find the annual temperature range.

Answer

$$\Delta T = T_{annual\ max} - T_{annual\ min} = 26\ °C - 1\ °C = 25\ °C$$

Do this one yourself!

What is the mean annual temperature for this year? Work out your response and then have a look. The answer is at the end of the unit.

Temperature Patterns

In order to clearly show how temperature varies at different locations, temperature maps are usually employed. A temperature map will consist of a base map of a particular region on which actual temperature measurements were taken at specific weather stations. These temperature readings are plotted on the map. The map is then further enriched by connecting all points that share the same temperatures using lines called **isotherms**. This way, the temperature pattern over a larger area can be discerned. A horizontal **temperature gradient** exists between high temperature regions and low temperature regions. Heat is transferred by advection from high temperature regions to low temperature regions. How quickly heat is transferred depends on how steep the temperature gradient is.

The map below (Figure 8.1) shows temperatures at various stations over an area. Notice how difficult it is to figure out the spatial pattern of temperature with only point measurements. To avoid this situation, you will need to add isotherms to this map. Follow the guidelines below to do this. Try it on your own before referring to the answer at the end of the unit.

Guidelines for Drawing Isotherms

1. Review the data on the weather map to decide how wide your isotherm interval will be.
2. Using a pencil, draw smooth isotherms on the map.
3. Locate the isotherms by interpolating between point measurements plotted on the weather map.
4. Isotherms cannot touch or cross.
5. Isotherms cannot branch or fork.
6. While being neat, ensure that you are as accurate as possible.

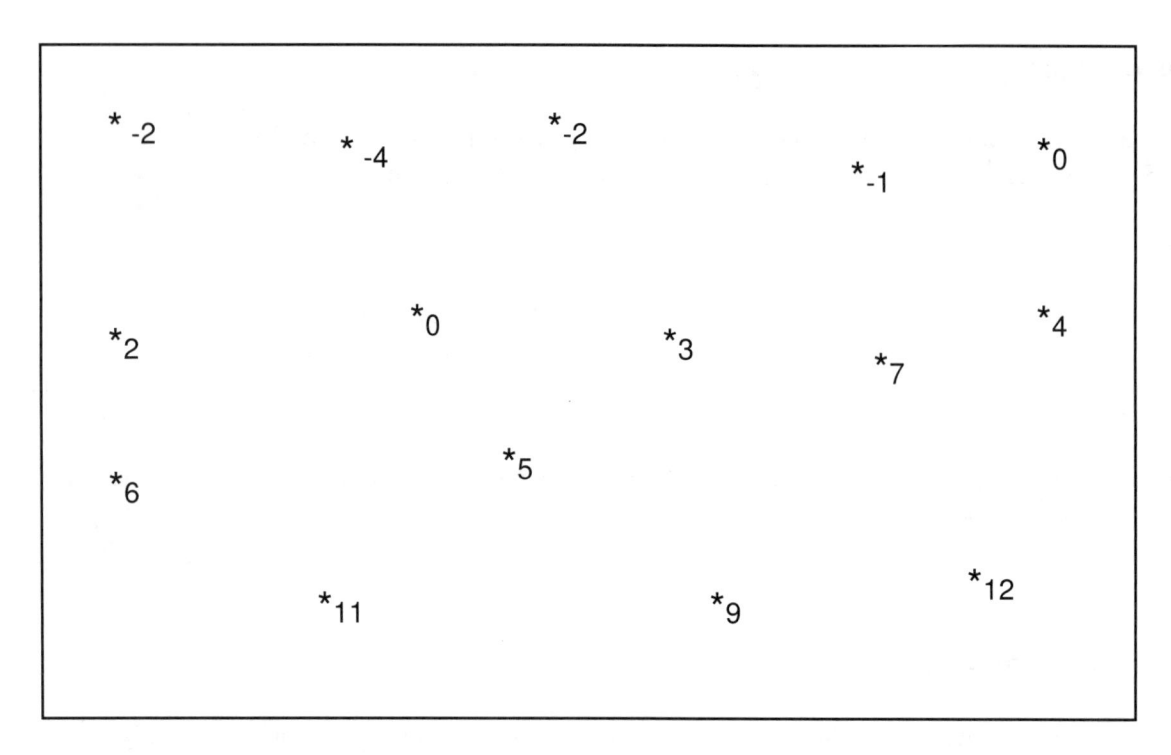

Figure 8.1. A map showing temperature readings over an area. The temperature is in degrees Celsius.

In the map below (Figure 8.2), you have been given a little help to start you off. Isotherms with an interval of 4°C have been partially inserted starting at -4°C, 0°C, and so on. Go ahead and extend these isotherms to the right side of the map. After isotherms are added, the task of interpreting the temperature pattern is much easier. It becomes obvious that temperature is reducing from bottom to top. Sensible heat is therefore transferred along the temperature gradient in that direction.

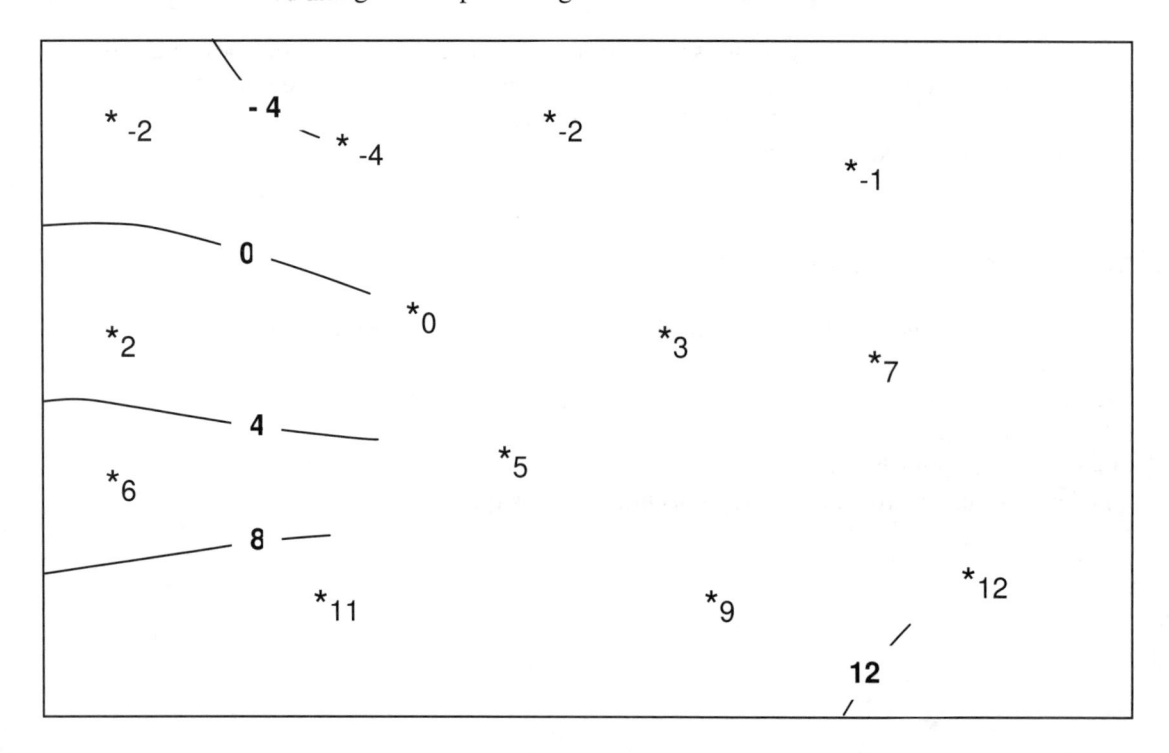

50

Solutions for Practice Questions

Mean daily temperature

$$\overline{T}_{daily} = \frac{T_{max} + T_{min}}{2} = \frac{12\,°C + -2\,°C}{2} = \frac{10\,°C}{2} = 5\,°C$$

Mean annual temperature

$$\overline{T}_{annual} = \frac{\Sigma \overline{T}_{monthly}}{12} = \frac{2+1+2+5+13+20+22+26+21+17+11+4}{12} = \frac{144}{12} = 12\,°C$$

Temperature map with completed isotherms.

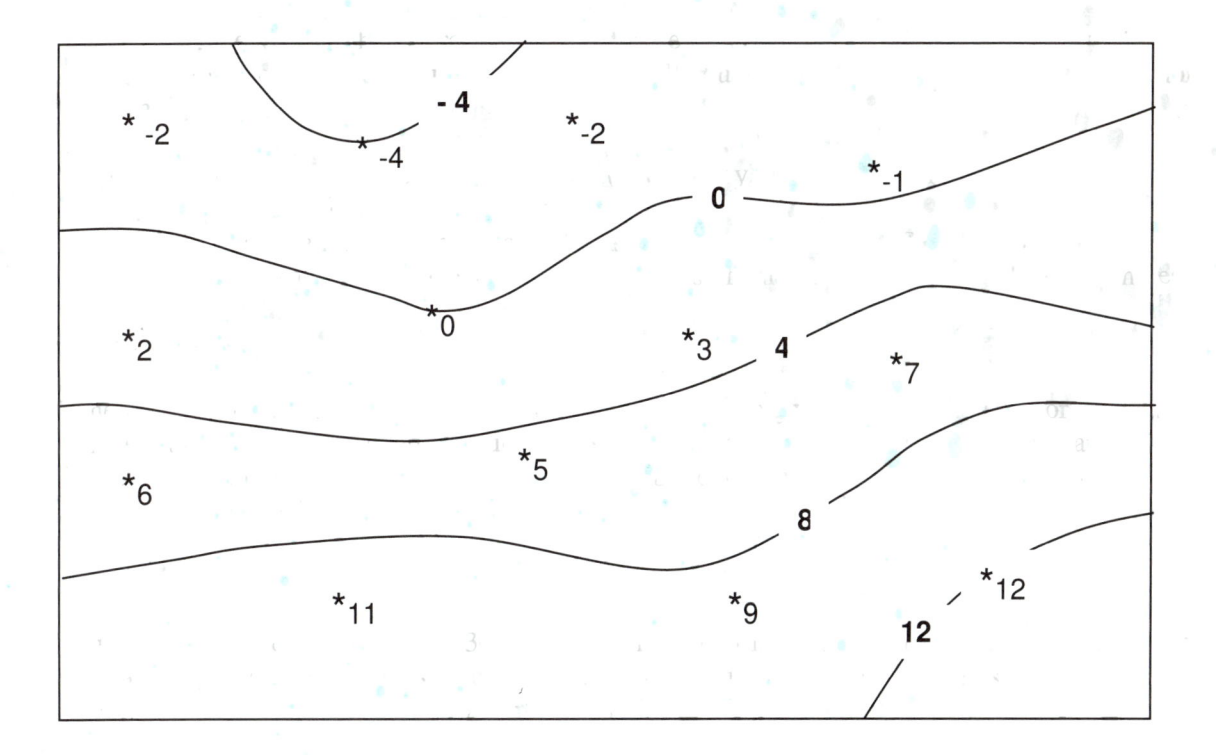

Air Pressure and Winds: A Supplement to Unit 9

This section discusses the following topics:

1. elevation and pressure
2. pressure gradient
3. Ideal Gas Law

At the end of this section, you will be able to convert station pressure to sea level equivalent pressure; to calculate the pressure gradient and estimate wind direction between two surface locations; and explain how a Galileo thermometer works.

Elevation and Pressure

By definition, pressure is force exerted per unit of surface area. Atmospheric pressure is basically the weight of air applied onto the Earth's surface. Force is exerted by the mass of air that is pulled down to Earth by gravity. We can express this relationship mathematically:

$$\text{Pressure} = \frac{\text{Weight} \times \text{gravity}}{\text{Area}} = \frac{\text{Force}}{\text{Area}}$$

This expression allows us to determine the units of pressure in the following manner:

$$\text{Units of pressure} = \frac{kg \times ms^{-2}}{m^2} = \frac{\text{Newtons}}{m^2} = \text{Pascals (Pa)}$$

It is clear from the above expressions that pressure depends a lot on the mass of air since gravity is assumed to be constant in the atmosphere. As you go up through the atmosphere, there will be less mass of air above you and therefore there will less pressure exerted on you by the atmosphere. This means that pressure reduces with increasing elevation in the atmosphere. From Figure 9.1, it is seen that the reduction in pressure in relation to elevation is exponential.

Although, this is the case, the rate at which pressure changes in the lower atmosphere is virtually constant. It reduces by 1 kPa for every 100 m that we ascend in the lower atmosphere. In order to have comparable readings of atmospheric pressure at different stations, we correct the values as if all stations were at sea level. We may calculate sea level pressure from station pressure at a known height as given below:

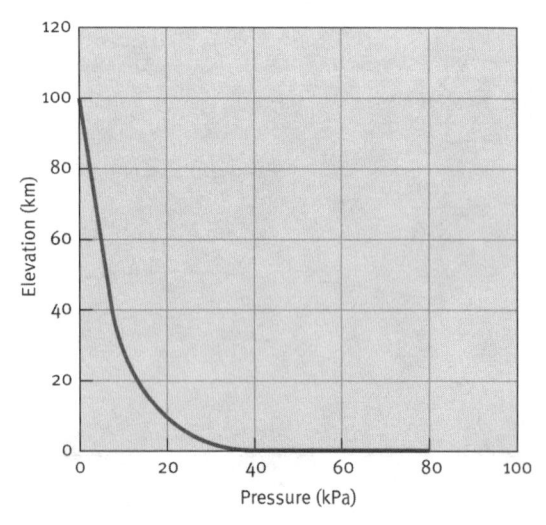

Figure 9.1 The relationship of pressure and elevation in a standard atmosphere.

Sea level pressure = station pressure + (station height × pressure change rate)

Example

What is the sea level pressure equivalent (P₀) for 88.6 kPa station pressure measured at Lusaka, Zambia which sits at 1154 m above sea level?

Answer

$$P_0 = 88.6\ kPa + \left(1154m \times \frac{1\ kPa}{100m}\right) = 88.6\ kPa + 11.54\ kPa = 100.14\ kPa$$

> **Now it's Your Turn**
>
> Another weather station at Mbala, Zambia is located at 1673m above sea level and has a pressure reading of 87.5 kPa. What is the sea level pressure equivalent? The answer is at the end of this unit.

Ideal Gas Law

Ideally, the pressure of a gas is dependent on the gas's density and temperature. The pressure P is greater where there are more molecules (i.e. greater density ρ) and where they are moving faster (i.e. greater temperature T). The law that determines this relationship is the Ideal Gas Law and is represented as follows:

$$P = \rho R.T$$

where R is the gas constant.

The Ideal Gas Law is well illustrated by the workings of the Galileo Thermometer (Figure 9.2). This is an inexpensive device made up of a closed glass cylinder containing water and a series of weighted glass balls. It is sealed so that the pressure inside cannot change. If the glass is heated with an infrared lamp for instance, the temperature goes up, and therefore, the density of the water must go down. In the formula, with pressure staying the same and temperature going up due to heating, the density must go down in order to satisfy the relationship in the Ideal Gas Law. As density decreases the glass balls sink. The lowest ball suspended in the cylinder gives the temperature of the fluid.

In the next unit, this law will help you understand how air rises and sinks in the atmosphere. It also helps to explain why thermal low-pressure systems have high temperatures and why thermal high-pressure systems have low temperatures.

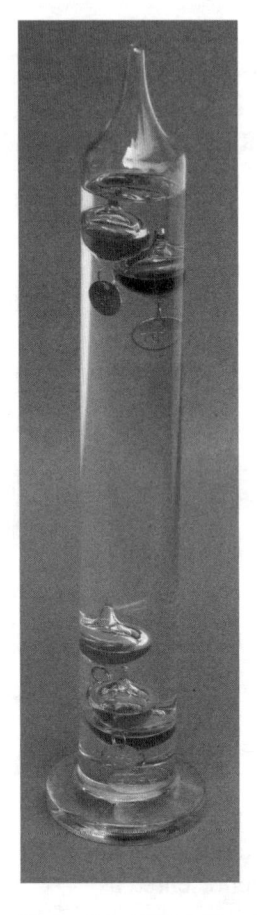

Figure 9.2. The Galileo thermometer shown on the left has two glass balls suspended at the top of the closed cylinder. The metal disk attached to the lowest of the suspended balls cites the temperature. When a hand is wrapped around the thermometer (middle picture) the fluid inside begins to heat up and become less dense. As a result, the next glass ball begins to descend (picture on the right). The new temperature will be given by the temperature stamped on the metal disk of the lowest glass ball that remains suspended.

Pressure Gradient and Wind Direction

As shown in the examples above, atmospheric pressure at different locations tends to be different. These differences in pressure create a **pressure gradient** (PG), which drives winds from a high-pressure system into a low-pressure system. The pressure gradient exerts a force (**pressure gradient force**, P) that acts perpendicular to **isobars**. Isobars are lines on a weather map that connect points that have the same atmospheric pressure. Winds that blow on Earth's surface depend on the strength of the pressure gradient force for their speed. If the pressure gradient is steep, the resultant pressure gradient force is equally large. The winds too will blow faster and with strength.

On weather maps, a large pressure gradient is depicted by isobars that are spaced very close to each other. This means that pressure changes quite rapidly over short distances on Earth. We are able to determine the pressure gradient that exists between any two locations on Earth where atmospheric pressure has been measured. The steps are as follows:

$$PG = \frac{\Delta P}{\Delta D}$$

where ΔP is the pressure difference and ΔD is the horizontal distance between two locations.

Because the Earth is continually spinning on its axis, winds are deflected by the Coriolis 'force' (C). Winds deflect to the right in the Northern Hemisphere and to the left in the Southern Hemisphere. This deflection is larger if the wind speed is high.

Due to the roughness of the Earth's surface, winds tend to be slowed down by the **friction force** (F), thus influencing the velocity and in turn the Coriolis 'force'. The **surface wind** (W) is a result of the interaction of all these forces and the associated environmental situations. We will illustrate this with an example of a couple of pressure systems in the Northern Hemisphere.

P = Pressure gradient force

C = Coriolis 'force' acting at 90° to the surface wind; the stronger the wind, the stronger the Coriolis 'force'

F = Friction force; depends on surface roughness

W = Surface wind

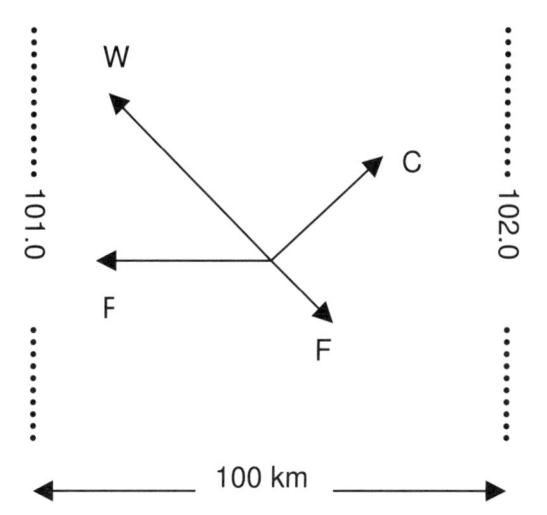

Figure 9.3 Wind flow in the area between two isobars in the Northern Hemisphere.

The pressure gradient in the example above is calculated as follows:

$$PG = \frac{102.0\ kPa - 101.0\ kPa}{100\ km} = 0.01\ kPa\ km^{-1}$$

Solution for Practice Question

Sea level pressure equivalent at Mbala, Zambia

$$P_0 = 85.5\ kPa + 1673m \times \frac{1\ kPa}{100m} = 85.5\ kPa + 16.73\ kPa = 102.23\ kPa$$

Circulation Patterns of the Atmosphere: A Supplement to Unit 10

The purpose of this section is to illustrate that the motion of winds in the atmosphere occurs at three basic scales and that pressure systems are not simply caused by heat differences.

Information is provided on:

1. three scales of atmospheric circulation:
    a. primary circulation
    b. secondary circulation
    c. tertiary circulation

2. factors inducing pressure systems:
    a. thermal low and high pressure systems
    b. dynamic low and high pressure systems

Primary Circulation

On a global scale, the general circulation pattern is regulated by large pressure systems. These include the Inter Tropical Convergence Zone (ITCZ), sub-tropical high-pressure cells, sub-polar low-pressure cells and polar high-pressure cells. The prevailing wind motion resulting from these pressure systems is called primary circulation.

An example of 1971–2000 normal westerly winds that blow all year round at Charlottetown, Prince Edward Island is given below. The data in the table below shows wind directions that are sustained for the entire year due to primary circulation. Although a slight difference appears in April and May, the rest of the year shows a westerly component in the winds.

Table 10.1. Normal wind direction at Charlottetown, Prince Edward Island. Source: Canadian Climate Data Centre and Archive

	Jan	Feb	Mar	Apr	May	Jun	Jul	Aug	Sep	Oct	Nov	Dec	Year
Wind Speed (km/h)	19.3	18.8	19.3	18.4	17.2	16.1	15.0	14.2	15.5	17.2	18.7	19.6	17.4
Most Frequent Direction	W	W	W	N	S	SW	SW	SW	SW	W	W	W	SW

Web activity

On the Canadian Climate Data Centre and Archive website (http://www.climate.weatheroffice.ec.gc.ca/climate_normals/index_e.html), find your own example of a location that shows very consistent normal wind directions for the entire year.

Secondary Circulation

At a regional scale, winds will be more variable and are classified as secondary circulation. A classic example of this type of circulation is seen in monsoon winds that change due to seasonal heating differences on water and land masses.

The figures below show wind roses at Darwin, Australia. A wind rose depicts the relative frequency of wind direction on an 8-point compass, with north, east, south, and west directions going clockwise. Each ring on the wind rose represents a frequency of 10 per cent of the total. You will note that in summer the wind blows from the North West and the West onto the land where there is a low-pressure system.

9 am Summer
4629 Total Observations
(1942 to 2004)

Calm 17%

Figure 10.1 The long-term wind rose for summer at Darwin, Australia. Source: Australian Bureau of Meteorology (http://www.bom.gov.au/climate/averages/wind/selection_map.shtml). Copyright Commonwealth of Australia, reproduced by permission.

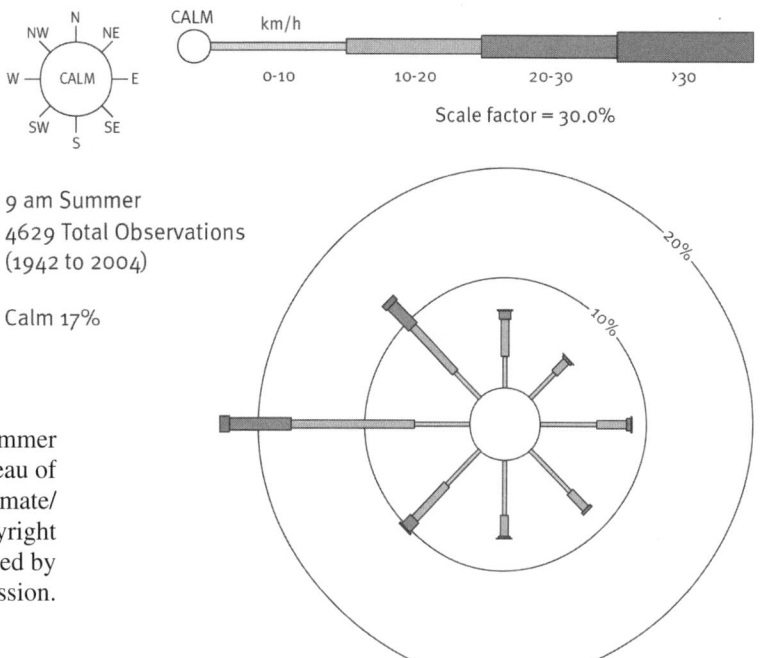

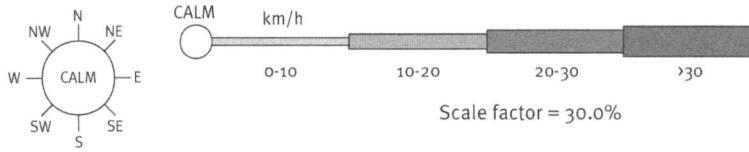

9 am Winter
4566 Total Observations
(1942 to 2004)

Calm 8%

On the other hand, winds reverse their direction and blow from East and South East onto to the ocean where there is a low-pressure system during winter.

Figure 10.2. The long-term wind rose for winter at Darwin, Australia. Source: Australian Bureau of Meteorology (http://www.bom.gov.au/climate/averages/wind/selection_map.shtml). Copyright Commonwealth of Australia, reproduced by permission.

The wind roses are read as follows:

Wind directions are divided into eight compass directions. The circles around the image represent the various percentages of occurrence of the winds. For example, if the branch to the west just reaches the 10 per cent ring it means a frequency of 10 per cent blowing from that direction. The scale factor can be ignored when interpreting these wind roses. An observed wind speed that falls precisely on the boundary between two divisions will be included in the lower range (e.g. 10km/h is included in the 1–10 km/h range). Calm has no direction.

Web activity

On the Australian Bureau of Meteorology website (http://www.bom.gov.au/climate/averages/wind/index.shtml), find a location that shows strong seasonal variation in wind direction.

Tertiary Circulation

At a much smaller scale, winds will change their direction more frequently depending on local physical factors. In Table 10.2, a sample of measurements of hourly wind direction at Abbotsford in British Columbia, Canada reveals that daytime sea breezes are common in summer. During day time, the land warms up quite rapidly in comparison to the Pacific waters in the west. A low-pressure system develops over the land and draws in wind from the Pacific and causes a sea breeze. Day-time heating is helped by long summer days. However, land breezes as seen in this case do not develop at all because the short summer nights do not permit the land's temperature to cool lower than the water temperature.

Date	Time	Temperature (°C)	Wind Direction (degrees)
August 25, 2002	0:00	16.2	
	1:00	15.3	
	2:00	14.6	
	3:00	14.8	320
	4:00	14.7	
	5:00	14.0	
	6:00	14.5	250
	7:00	16.5	250
	8:00	17.7	260
	9:00	17.8	260
	10:00	19.5	240
	11:00	19.0	230
	12:00	19.4	250
	13:00	20.6	270
	14:00	22.2	250
	15:00	21.5	230
	16:00	20.7	210
	17:00	20.1	240
	18:00	19.7	260
	19:00	18.6	220
	20:00	17.7	250
	21:00	16.9	240
	22:00	16.2	
	23:00	15.5	
August 26, 2002	0:00	15.7	
	1:00	15.7	
	2:00	15.7	310
	3:00	15.8	
	4:00	15.7	270
	5:00	15.2	
	6:00	15.7	
	7:00	16.3	
	8:00	16.9	
	9:00	18.7	
	10:00	20.0	280
	11:00	21.7	
	12:00	22.8	270
	13:00	23.5	240
	14:00	24.6	210
	15:00	24.7	220
	16:00	24.3	240
	17:00	24.4	260
	18:00	22.6	260
	19:00	20.8	270
	20:00	18.9	270
	21:00	17.4	
	22:00	15.6	
	23:00	15.0	

Table 10.2. Hourly temperature and wind direction at Abbotsford, British Columbia. Source: Canadian Climate Data Centre and Archive.

Thermal and Dynamic Pressure Systems

Some low-pressure systems result from a net radiation surplus. As the air heats and rises, horizontal winds are brought in to fill the gap that is created inside the **thermal low**. This is typically the case along the equatorial region where all year round high temperatures create the ITCZ. On the contrary, a net radiation deficit exists at the poles and the air there is cold and dense. This air subsides and causes a **thermal high** at the surface. You will recall from Unit 9 how the Ideal Gas Law underlies the workings of the Galileo Thermometer; this is also applicable to the formation of thermal low and high pressure systems.

Other pressure systems are not created by net radiation, but rather by motion in the upper atmosphere. Convergence of upper winds forces some air to subside downward and create what is known as a **dynamic high** at the surface. If upper winds tend to diverge, they will force air below to rise and form a **dynamic low** at the surface. We will summarize how primary pressure systems develop and describe their associated properties in the table below.

Table 10.3. Primary pressure systems and their properties

Name	Cause	Location	Air Temperature	Humidity	Atmosphere
Polar high-pressure cells	Thermal	90° N; 90° S	Cold	Dry	Stable
Subpolar low-pressure cells	Dynamic	60° N; 60° S	Cool	Wet	Unstable
Subtropical high-pressure cells	Dynamic	20°–35° N and S	Hot	Dry	Stable
Equatorial low-pressure trough	Thermal	10° N–10° S	Warm	Wet	Unstable

We may be familiar with high-pressure systems that occur in summer in mid-latitude regions. Although the air is warm, it does not rise upward but instead it sinks to the surface and causes a high-pressure system. Such a system is a dynamic high. This is usually accompanied by clear skies and a lack of precipitation. Remember that if the system were thermal, the warm air would have risen. Also we may be familiar with low-pressure systems that develop in cold winter. In these systems, relatively cold air rises upward and causes stormy weather. This is a good example of a dynamic low. If this were an ordinary thermal system, the relatively cold air would have had no ability to rise upward. The air would settle on the ground and the weather would have been less stormy. Under these two situations that we have discussed, the Ideal Gas Law is not responsible for convecting air into vertical motion.

Moisture in the Atmosphere: A Supplement to Unit 12

This section is designed to supplement and expand on material first introduced in the text in Unit 8 (Temperature). These concepts are examined here because they are particularly relevant to an understanding of both moisture in the atmosphere and adiabatic processes that form clouds and possibly precipitation.

Measures of Moisture in the Atmosphere

We can express the amount of moisture contained in the atmosphere in several ways depending on what use we have for moisture content. Most people choose to use **vapour pressure** and **relative humidity** to quantify atmospheric moisture. If an air parcel holds the maximum amount of water that it can contain at it's current temperature, it is said to be saturated. We can bring an unsaturated air parcel to saturation by either adding moisture to it or by cooling it so that its capacity to hold moisture lowers. The critical point at which air is saturated is called the **dew point temperature** and is associated with the maximum moisture content.

Vapour pressure

The weight of each gas in the atmosphere contributes to the total atmospheric pressure and so does water vapour. The **actual vapour pressure** of air refers to the pressure caused by water vapour and thus provides us with a measure of moisture contained in the air. When an air parcel is saturated, its pressure is described as the **saturation vapour pressure.** Note that saturation vapour pressure is directly proportional to air temperature.

The saturation vapour pressure of air (e_s) measured in millibars as a function of air temperature (°C) is presented in Table 12.1 in the range -40°C to 40°C. The saturation vapour pressure for air at even-numbered temperatures can be read straight off the table, while e_s for air at temperatures between these given levels must be interpolated using the average rate of change values listed in the table. Two examples will illustrate the procedure

1. For a temperature of -18° C, the saturation vapour pressure of air can easily be read off as 1.49 mb. This value was read right off the table.

2. Yet the saturation vapour pressure associated with air at 10.6°C can be a little difficult to arrive at. Since this temperature lies within a range on the table, its saturation vapour pressure must be between 12.27 mb and 14.02 mb. The exact saturation vapour pressure will be the value for 10°C (12.27 mb), plus the difference between 10°C and 10.6°C (0.6°C) multiplied by the rate of change for this interval (0.875mb °C^{-1}). Thus, e_s for air at 10.6°C = 12.27 mb + (0.6 mb)(0.875mb °C^{-1}) = 12.795 mb.

Table 12.1. Saturation vapour pressure e_s (mb) for selected temperatures (°C)

T(°C)	e_s (mb)	Δe_s (mb °C^{-1})	T(°C)	e_s (mb)	Δe_s (mb °C^{-1})
-40	0.189		0	6.11	
		0.022			0.470
-38	0.232		2	7.05	
		0.026			0.540
-36	0.284		4	8.13	
		0.031			0.610
-34	0.346		6	9.35	
		0.037			0.685
-32	0.420		8	10.72	
		0.045			0.775
-30	0.509		10	12.27	
		0.052			0.875
-28	0.613		12	14.02	
		0.062			0.975
-26	0.737		14	15.97	
		0.073			1.100
-24	0.883		16	18.17	
		0.084			1.230
-22	1.05		18	20.63	
		0.100			1.370
-20	1.25		20	23.37	
		0.120			1.530
-18	1.49		22	26.43	
		0.135			1.700
-16	1.76		24	29.83	
		0.160			1.890
-14	2.08		26	33.61	
		0.180			2.090
-12	2.44		28	37.79	
		0.210			2.320
-10	2.86		30	42.43	
		0.245			2.560
-8	3.35		32	47.55	
		0.280			2.825
-6	3.91		34	53.20	
		0.315			3.110
-4	4.54		36	59.42	
		0.365			3.420
-2	5.17		38	66.26	
		0.420			3.755
0	6.11		40	73.77	

Δe_s shows the average rate of change of saturation vapour pressure within one temperature interval in the table. This average rate can thus be employed to estimate saturation vapour pressures for temperatures that lie anywhere in the interval. Suppose you are interested in determining the e_s for air whose temperature is 36.6°C. The air being between 36°C and 38°C, its e_s will be the value for 36°C (59.42 mb) plus the difference in the temperature from the lower boundary of the interval (0.6°C) x the average rate of change for this interval (3.420 mb °C^{-1}).

$$\text{Answer} = 59.42 + (0.6°C)(3.420 \text{ mb °C}^{-1}) = 61.472 \text{ mb.}$$

Humidity

Although humidity is a general term describing the water vapour contained in air, there are various indices used to assess this quantity. The most frequently used measure in every day life is **relative humidity**. This expresses the proportion of the content of moisture in relation to the capacity of air to hold moisture. In other words, relative humidity (RH) may be defined as the actual vapour pressure (e_a) of the air as a percentage of the saturation vapour pressure (e_s).

$$RH = \frac{e_a}{e_s} x100$$

Example: If an air parcel has actual vapour pressure of 9 mb and saturation vapour pressure of 18 mb, then the relative humidity is:

$$RH = \frac{9mb}{18mb} x100 = 50\%$$

Dew point temperature

The **dew point temperature** (T_d) is the temperature at which an air parcel would become saturated if it were cooled at constant pressure or constant moisture content. This refers to the temperature at which the air would become saturated strictly due to cooling. When the air cools beyond the dew point, water vapour in the air will be condensed into water droplets. The difference between the air temperature and the dew point temperature is called the dew point depression. The larger the dew point depression, the drier the air.

Equipped with Figure 12.1, we are able to determine the dew point that corresponds with any given saturation vapour pressure in the range -40°C to + 40°C. Graphically we can tell that when the saturation vapour pressure is 30 mb, the dew point will be close to 24°C. This case applies to an air parcel that is already saturated. If the air parcel is unsaturated and its temperature is 24°C, we may say that its actual vapour pressure should be increased to 30 mb in order to be saturated. **Note**: At saturation, $e_a = e_s$ and $T_a = T_d$.

Influence of Temperature on Saturation Vapour Pressure

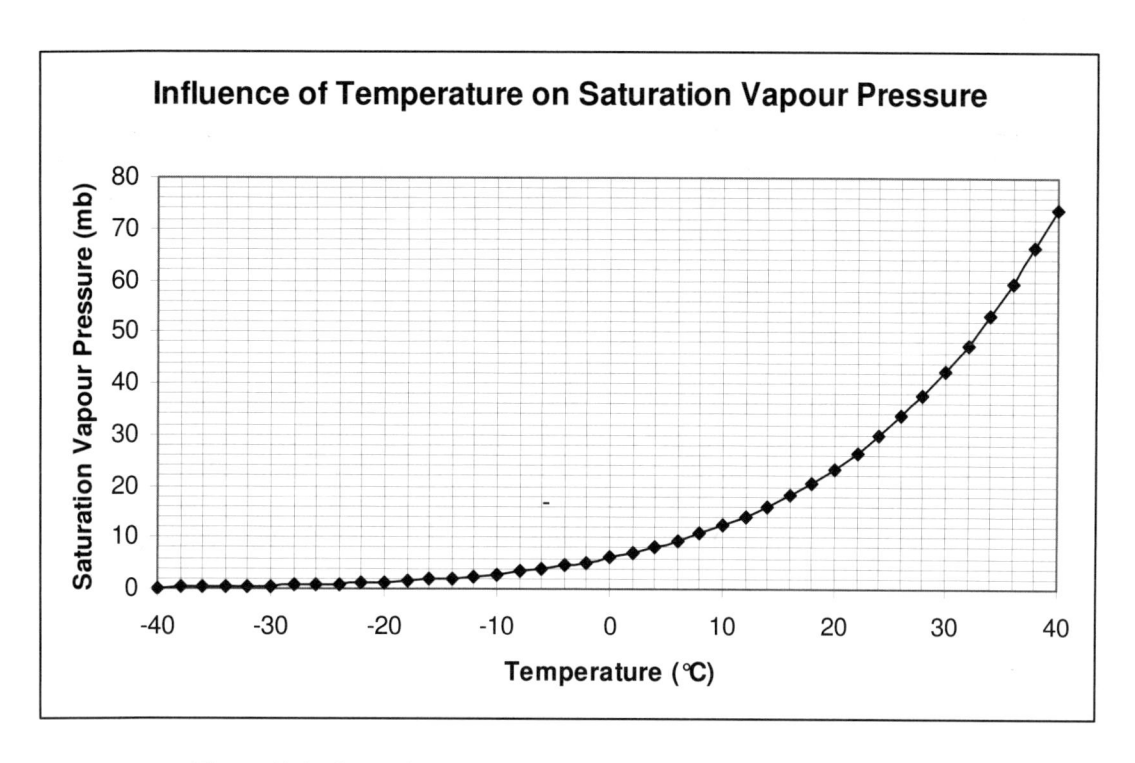

Figure 12.1. Saturation vapour pressure as a function of air temperature.

We will work with examples to fill in the following table:

Air Temperature (°C)	Saturation Vapour Pressure, e_s (mb)	Dew Point Temperature, T_d (°C)	Actual Vapour Pressure, e (mb)	Relative Humidity (%)
24			5	

1. Given an air temperature of 24°C, from table 12.1 we get a saturation vapour pressure of 29.83 mb. Fill this into the table above.

2. Given an actual vapour pressure of 5 mb and a saturation vapour pressure of 29.83 mb, we can calculate the relative humidity.

$$RH = \frac{5mb}{29.83mb} \times 100 = 16.8\%$$

Again, fill this in on the table above.

3. We then use Figure 12.1 to find the dew point temperature of this air if it cooled at constant moisture content. With actual vapour pressure at 5 mb, the corresponding temperature is approximately -4°C. The air must cool to -4°C in order to saturate and condense. Since the dew point depression is large, this air parcel is obviously very dry.

Adiabatic Processes and Stability

Ordinarily, air temperature will change when sensible heat is added to or removed from the air. It is also possible for an air parcel to experience a temperature change without gaining heat from or losing heat to the surrounding environment. The latter case is described as an **adiabatic process**. When air rises, it encounters an atmosphere with less pressure and expands outward and cools. Similarly, when air subsides, it comes into an atmosphere that has greater pressure and is compressed and heated.

Air that has RH < 100 per cent is unsaturated air and will rise and cool, or sink and warm, at the constant dry adiabatic lapse rate (DAR) of 10°C km^{-1}. If air has RH ≥ 100 per cent, it is saturated and its adiabatic process will also involve condensation or evaporation of moisture within the air parcel. The latent heat associated with this phase in a saturated parcel will offset the dry adiabatic rate by some amount. This leads to a lower moist adiabatic lapse rate (MAR) of approximately 6°C km^{-1}. When a saturated air parcel cools adiabatically, some of the water vapour in the parcel condenses and releases latent heat that partially warms the air parcel. Conversely, when a saturated air parcel warms adiabatically, some of the water droplets in the air parcel will evaporate and use up latent heat, which cools the air parcel by some degree.

Based on these adiabatic lapse rates and the environmental temperature profile, we may deduce the air's tendency to rise or not to rise (i.e. its **stability**). To understand this, consider the example in the figure below. The boulder settled in the depression is in a state of STABLE equilibrium such that if it were pushed upslope it will tend to roll back to the depression bottom. Another boulder perched on the summit of a mound is in UNSTABLE equilibrium, as it will accelerate away from the summit if it was tapped over. But a boulder on flat ground will move only when it is pushed and will remain at the new location after the force is released. Such a situation where there is no tendency to continue moving away from or to return to the original position is described as NEUTRAL equilibrium.

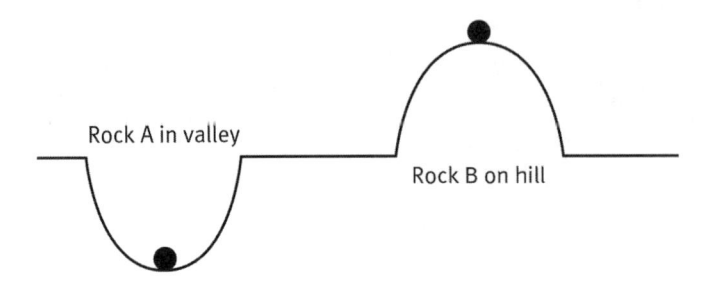

If air is in a state of stable equilibrium, it resists all upward or downward motion by returning to its initial position. If an unstable equilibrium prevails in the atmosphere, convection is readily possible as the air can rise or sink freely. In some cases, an air parcel will exist in neutral equilibrium and will tend neither to sink, rise nor return to its original position.

Graphs can be very helpful in determining the stability of the atmosphere by comparing the adiabatic process to the surroundings of an air parcel. How the temperature in the environment surrounding the parcel changes is termed the environmental lapse rate (ELR). If a rising air parcel becomes colder than the environment, it will tend to sink back because it is denser than its surroundings. This is an example of **absolute stability**. If a rising air parcel is warmer than the environment, it will continue rising because it is less dense than its surroundings. This is an example of **absolute instability**. But air that comes into an environment that has the same temperature will come to rest. This is an example of **neutral stability.**

Example

A meteorological balloon is sent aloft and the following data are obtained:

Table 12.2. Sounding of environmental temperature in the atmosphere

Height (m)	Temperature(°C)	Height (m)	Temperature(°C)
0	20.2	625	17.2
125	19.9	775	17.1
250	19.4	900	16.9
375	18.8	975	17.2
500	18.1	1200	17.4

These data are plotted on the following figure to illustrate the environmental lapse rate. A 'hotspot' of air at surface with a temperature of 24°C and dew point temperature of 19°C rises through this environment. Its path as it rises is plotted on the graph (Figure 12.2).

Before saturation, the parcel follows the DAR. When it cools to the dew point it then switches to the MAR. It is seen on the graph that the air is unstable from the surface until a height of about 800 m it becomes neutral. If the air parcel rises beyond this level, it will be colder than its surroundings and will sink back to this level. So above the 800 m height the atmosphere is stable.

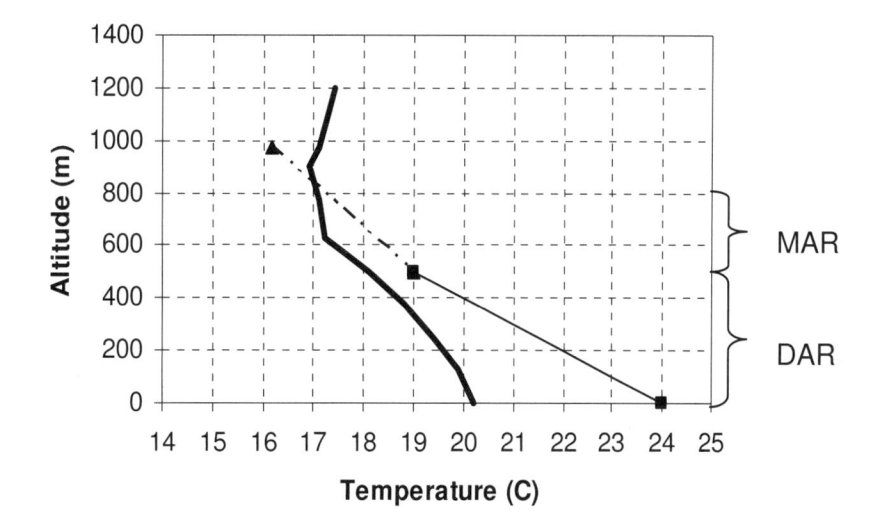

Figure 12.2. A graph showing the environmental temperature and the temperature of an air parcel that is cooling adiabatically. The thick solid line represents the environmental temperature; the thin solid line represents the air parcel's temperature before saturation; and the dashed line represents the sir parcel's temperature after saturation.

Example

Warm, moist air with a sea level temperature of 15°C is forced to rise over the Coast Mountains towards Eastern Canada. Its dew point temperature is measured as 10°C. If the summit of the mountains is 1,000 m, what is the temperature of the air parcel when it reaches there? As the air parcel sinks on the leeward side, what will its temperature be at 300 m?

Applying the DAR since the air parcel is, as yet, unsaturated shows that saturation will occur at the condensation level given by:

$$5°C\left(\frac{1000m}{10°C}\right) = \frac{5000°Cm}{10°C} = 500m$$

Then the air parcel will start cooling at the MAR because condensation is now occurring inside it and latent heat is issued. The air parcel will ascend another 500 m to get to the summit of the mountains and the air parcel will cool by:

$$500m\left(\frac{6°C}{1000m}\right) = \frac{3000°Cm}{1000m} = 3°C$$

The new temperature at the summit is 10°C - 3°C = 7°C.

On the leeward slope, the parcel descends through 700m at the DAR. It warms by:

$$700m\left(\frac{10°C}{1000m}\right) = \frac{7000°Cm}{1000m} = 7°C$$

So its final temperature is 7°C + 7°C = 14°C.

Minerals and Igneous Rocks: A Supplement to Unit 30

You may be enrolled in a full year of physical geography. If you are, your professor or instructor may introduce you to more minerals and rocks than you might encounter in a one-semester program. This supplementary unit is meant to provide additional information for Unit 30 of *Physical Geography* to complement mineral and igneous rock laboratory assignments.

Minerals

Having read Unit 30 in the text, you know that minerals are the building blocks of rocks, and therefore, are a perfect launching pad for you to begin constructing your knowledge of Earth's surface materials.

Imagine this scenario. . . a specific mineral is placed in front of you during your lab class, you are given hundreds of pages of descriptions for the few thousand minerals that exist on planet Earth, and you are asked to identify this particular mineral. You pick up the pages of information, thinking, 'Okay, how hard can this be?', and quickly realize that the information is organized by mineral name in alphabetical order. Where should you start? If you start in the 'A's and your mineral begins with a 'P' you'll be there for days, becoming more and more frustrated and confused. If you skip over pages, you could miss the description of your mineral and have to start all over again. What you need is information that is categorized and organized to simplify the identification process!

Happily for us, geologists have already classified minerals according to a few straight-forward criteria linked directly to each mineral's unique properties. All you need to do to begin is to ask a few simple questions:

- Does this mineral have a metallic lustre (like a shiny, new metal coin or an old, dull coin)?
- Is this mineral light or dark in colour?
- Is it harder than glass (does it scratch glass)?
- Does it have cleavage (does it break into a series of smooth flat surfaces at the same angles)?

Lustre

If the lustre is non-metallic, various other descriptors could be given, including: *glassy* or *vitreous* if the surface reflects light like clear or coloured glass does; *adamantine*, if it reflects light brilliantly like diamond does; or *resinous*, if it reflects light like amber does. Other lustres could be *pearly*, *silky*, or *greasy*, which are self-explanatory.

Colour

While you know from reading Unit 30 in the text that colour is not a particularly reliable property of minerals, it does serve to help us quickly classify minerals into categories. We'll use it here to simply indicate whether minerals are light or dark in colour.

Hardness

To test how hard the mineral is we compare it to the hardness of a glass plate, which measures between 5 and 6 on the Moh's Hardness Scale (Table 30.1 from the text). To do so, take a rectangular glass plate supplied in your lab and lay it down flat on your desk. Find a fresh corner of your mineral sample and draw it firmly across the plate. Check to see whether your mineral scratched the plate or not. If it did, the mineral is harder than glass, so has a hardness of 6 or more.

Later on, when we're coming close to identifying the mineral, we will fine-tune the hardness measurement by scratching it against other objects. If it is softer than glass, we can compare it to the hardness of a copper penny (3.5 on Moh's Scale) or our fingernails (2.5). If it is harder than glass, we can make a relative comparison by scratching it against another hard mineral to find out which is harder and which is softer.

Table 30.1 The Mohs' Hardness Scale

Mineral	Hardness
Diamond	10
Corundum	9
Topaz	8
Quartz	7
Potassium feldspar	6 — Glass
Apatite	5 Plate
Fluorite	4
Calcite	3
Gypsum	2
Talc	1

By way of comparison, here are some everyday items ranked according to their approximate hardness: pocketknife blade, 5–6; glass, 5; copper penny, 3.5; fingernail, 2.5.

Cleavage

Identifying the cleavage of a mineral can be challenging. The cleavage represents the tendency of a mineral to break along consistent lines or cleavage planes. It is associated with weak bonding between atoms in particular directions. In Figure 30.1a, for instance, the bonds between atoms are strong in the vertical direction, but weaker in the horizontal direction. If we take a hammer and strike a sample of this mineral, it will break along the lines of weak bonding. It will break in the directions indicated by the dotted lines. The mineral illustrated in Figure 30.1 has one direction of **cleavage** (1-D cleavage).

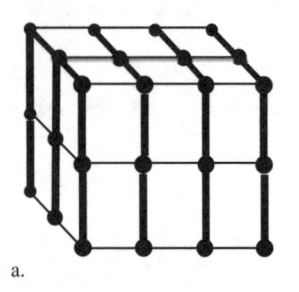

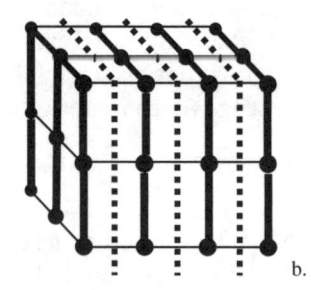

Figure 30.1 Diagrams of a mineral sample before and during breakage (a. and b. respectively). In Fig. 30.1a strong bonds between atoms are illustrated as wider connecting lines while weaker bonds are shown with thinner lines. When subjected to stress, the mineral will break consistently along the lines of weak bonds as shown with the dotted lines in Fig. 30.1b.

In Figure 30.2, a different situation is shown. Here the bonds are equally strong in all directions. If we strike a sample of this mineral with a hammer, it will break randomly. Instead of cleaving, it will **fracture** into an irregular surface.

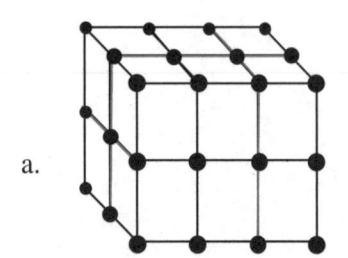

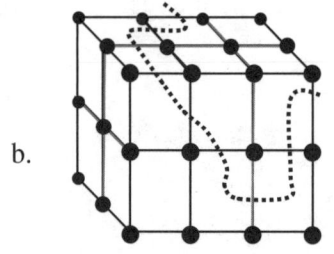

Figure 30.2 In this mineral, all bonds are of the same strength (Fig. 30.2a). When it is subjected to stress, it will break randomly, in other words it will **fracture** into an irregular surface (Fig. 30.2b).

For our purposes, we will recognize that a mineral either cleaves *or* fractures; it does not do both. This is true for most minerals.

Figure 30.3 illustrates the main types of cleavage in minerals. Cleavage in one direction produces minerals that break into thin sheets. This type of cleavage is easy to recognize in the family of mica minerals, for instance.

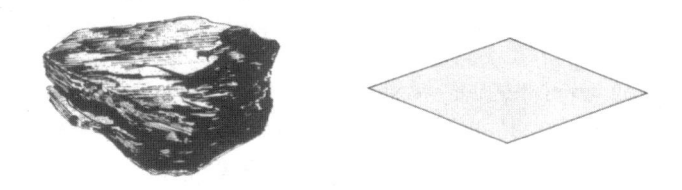

A. CLEAVAGE IN ONE DIRECTION. Example: muscovite.

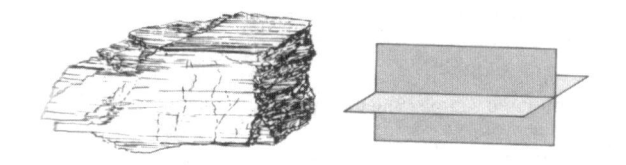

B. CLEAVAGE IN TWO DIRECTIONS AT RIGHT ANGLES. Example: feldspar.

C. CLEAVAGE IN TWO DIRECTIONS NOT AT RIGHT ANGLES. Example: amphibole.

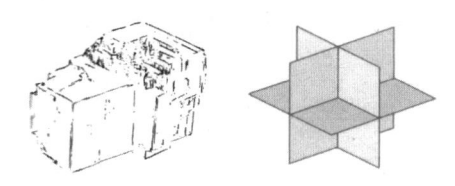

D. CLEAVAGE IN THREE DIRECTIONS AT RIGHT ANGLES. Example: halite.

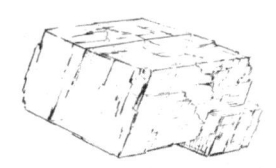

E. CLEAVAGE IN THREE DIRECTIONS NOT AT RIGHT ANGLES. Example: calcite.

Figure 30.3   Common types of cleavages and minerals that express them. From Hamblin, W.K. and J. D. Howard. *Exercises in Physical Geology*, 12th Edition, © 2005, p. 11. Reprinted by permission of Pearson Education, Inc., Upper Saddle River, NJ.

Cleavage in two directions produces a tiny 'staircase' of risers and treads. All the risers are at the same angle, and all the treads are at a different, but consistent angle. These risers and treads may only be a millimetre or less across, so they are not easily seen without a binocular microscope. You might see (or feel!) 2-D cleavage on the outer edges of your mineral sample, but more likely, you will find it as a subtle staircase down the faces of the samples.

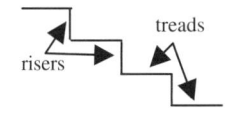

Figure 30.4   2-D cleavage produces a tiny staircase that might be felt more easily than seen with the naked eye.

If you think you have 2-D cleavage, become tactile and treat the sample as braille. Rotate its faces in your hand; rub your thumb along them. Is a staircase present? If one is, you have 2-D cleavage. If you want to give a more sophisticated analysis, try to determine whether the risers and treads are perpendicular to one another. You could do this by looking at your sample under a point light source.

If you slowly rotate the sample in your hands, you will see many tiny cleavage planes catch the light at the same time, indicating that they are all at the same angle. Rotate the sample again, until you see a set of different tiny cleavage planes pick up the light. Now think about how far you have rotated the sample. Has it been 90°? If you have a hand lens or a binocular microscope present in your classroom, check your sample out under magnification to confirm your suspicions about the type of 2-D cleavage you have.

Cleavage in three directions is generally clear to see. If you have 3-D cleavage at right angles, cubes will be formed. If the cleavage is 3-D, but not at 90°, then a rhombohedral (a skewed cube) is produced. Some of the common minerals that display 3-D cleavage are translucent. If you look at them closely you may be able to see internal fracture planes that indicate future cleavage planes along which the mineral will break.

Figure 30.5. 3-D cleavage at right angles (**cubic**) on the left, and not at right angles (**rhombic**) on the right. The thin lines within the rhombic sample illustrate internal fracture planes along which the sample will break in the future.

Fracture

If a mineral does not cleave, then for our purposes in first-year physical geography, it **fractures** into an irregular surface. This means that the bonds between atoms are equally strong in all directions, so there are no preferred lines of breakage. When you hold a mineral that fractures up to the light, some facets will reflect light regardless of the angle you hold the sample at.

Mineral Identification

You are almost ready to begin mineral identification. Two further pieces of information will be helpful. First, if a mineral solidifies in a confined space, its crystals jumble together as they grow into each other, giving the sample a *massive habit*. If the same mineral grows in unconfined space, it may express *crystal faces*—flat, smooth faces that are not cleavage planes, but rather, grow naturally in unconfined space. To keep things simple, *crystal habit* information is included in the following minerals keys only for minerals that express clear crystal faces at specific angles.

Second, some minerals have a hardness value very similar to that of glass. Some samples might scratch glass while others would not. To avoid confusion, those minerals are listed under both hard and soft minerals. Similarly, because colour varies in some minerals, those that may have hard-to-define colours are listed under both light and dark sections.

Are you ready to begin mineral identification? Pick a sample and ask yourself these questions.

Question 1: **Does the mineral have a metallic lustre?**
Yes – go directly to Table 30.2
No – go to Question 2

Table 30.2 Some Common Metallic and Submetallic Minerals

Streak	Lustre, Colour and Streak	Cleavage or Fracture	Crystal Habit and Other Notes	Hardness	Name and Chemical Composition
RED STREAK	Lustre: can be dull or bright Colour: steel grey, red grey, to black Streak: red-brown	Cleavage: none Fracture: uneven	Crystal Habit*: may exhibit botryoidal (grape-like cluster) shapes	5-6	**HEMATITE** Fe_2O_3
YELLOW, BROWN STREAK	Lustre: dull, earthy Colour: yellow, yellow brown, or black Streak: yellow, brown or black	Cleavage: perfect, but not readily visible	Other Notes: bog iron ore	5-5.5	**LIMONITE** $Fe_2O_3 \bullet H_2O$
DARK GREEN, GREEN BLACK, OR BLACK STREAK	Lustre: often dull Colour: dark grey to black Streak: black	Cleavage: none Fracture: uneven	Other Notes: strongly magnetic	6	**MAGNETITE** Fe_3O_4
	Lustre: bright Streak: lead grey to grey-black Colour: lead grey	Cleavage: cubic; 3-D at 90°	Crystal Habit*: cubic crystals Other Notes: wash your hands after handling lead-rich galena	2.5	**GALENA** PbS
	Lustre: dull Colour: steel grey to black Streak: black	Cleavage: 1-D, although not often visible in soft, handled samples	Other Notes: feels greasy, marks paper	1-2	**GRAPHITE** C
	Lustre: bright Colour: pale brass yellow Streak: green-black	Cleavage: none Fracture: conchoidal-uneven	Crystal Habit*: cubic crystals often with striated faces Other Notes: "Fool's Gold"	6-6.5	**PYRITE** FeS_2
	Lustre: metallic Colour: brass yellow, may tarnish to bronze or purple Streak: greenish black	Cleavage: none Fracture: uneven		3.5-4	**CHALCOPYRITE** $CuFeS_2$

*Remember that the crystal habit will only be expressed if the sample has grown in unconfined space; most samples will not exhibit this.

Question 2: **For non-metallic minerals, is the mineral light- or dark-coloured?**
      Light-coloured – go to Table 30.3.
      Dark-coloured – go to Table 30.4.

Question 3: **Does the mineral scratch glass?**
      Yes – go to the top half of the tables
      No – go to the bottom half of the tables

Table 30.3 Some Common Non-Metallic, Light-Coloured Minerals

		Lustre, Colour, and Streak	Cleavage or Fracture	Crystal Habit and Other Notes	Hardness	Name and Chemical Composition
HARD (SCRATCHES GLASS)	**CLEAVAGE**	Lustre: vitreous to pearly Colour: varies from colourless/white to brown, commonly salmon-coloured Streak: white or colourless	Cleavage: two planes at nearly right angles	Other Notes: lack striations seen in other feldspars	6	**ORTHOCLASE (K-FELDSPAR)** $KAlSi_3O_8$
		Lustre: vitreous to pearly Colour: grey-white Streak: white or colourless	Cleavage: two planes at close to right angles	Other Notes: striations common on good cleavage surfaces	6-6.5	**Na-RICH PLAGIOCLASE FELDSPAR** $NaAlSi_3O_8$ (Albite)
	NO CLEAVAGE	Lustre: vitreous Colour: colourless; white; pink; purple; grey; any colour may result from impurities Streak: colourless	Cleavage: none Fracture: conchoidal	Crystal Habit*: prismatic crystals with striations perpendicular to the long dimension	7	**QUARTZ** SiO_2
		Lustre: vitreous Colour: light to dark olive green Streak: pale green to white	Cleavage: none Fracture: conchoidal, but not well expressed in small samples		6.5-7	**OLIVINE** $(Mg, Fe)_2SiO_4$
		Lustre: adamantine to vitreous on unweathered surfaces Colour: various colours, usually brown, grey, pink, or blue; typically dark, but sometime lighter Streak: harder than a streak plate – no streak	Cleavage: none Fracture: breaks easily across crystal	Crystal Habit*: six-sided, barrel-shaped crystals, sometimes with deep horizontal striations	9	**CORUNDUM** Al_2O_3
SOFT (DOES NOT SCRATCH GLASS)	**CLEAVAGE**	Lustre: vitreous Colour: colourless or white; other colours due to impurities Streak: white or colourless	Cleavage: cubic; 3-D at right angles	Other Notes: salty taste	2.5	**HALITE** NaCl
		Lustre: vitreous to earthy Colour: colourless, white, yellow Streak: white or colourless	Cleavage: rhombohedral (3-D, not at 90°)	Other Notes: effervesces in cold dilute HCl	3	**CALCITE** $CaCO_3$
		Lustre: vitreous to silky or pearly Colour: colourless to pale green/brown Streak: white or colourless to light yellow/brown	Cleavage: 1-D; splits into thin elastic sheets		2-2.5	**MUSCOVITE** $KAl_3Si_3O_{10}(OH)_2$
		Lustre: pearly to greasy Colour: green to white or silver white Streak: white or colourless	Cleavage: 1-D	Crystal Habit*: crystals extremely rare Other Notes: soapy feel	1.0	**TALC** $Mg_3Si_4O_{10}(OH)_2$
		Lustre: vitreous or pearly Colour: White, colourless and light shades of red, blue and yellow Streak: white or colourless	Cleavage: 2-D, perfect in one direction	Crystal Habit*: often groups of tabular crystals	3-3.5	**BARITE** $BaSO_4$
	NO CLEAVAGE	Lustre: resinous Colour: light yellow Streak: yellow	Cleavage: none Fracture: conchoidal to uneven	Other Notes: smells of sulphur, or rotten eggs	2	**SULPHUR** S

*Remember that the crystal habit will only be expressed if the sample has grown in unconfined space; most samples will not exhibit this.

Table 30.4 Some Common Non-Metallic, Dark-Coloured Minerals

	Lustre, Colour, and Streak	Cleavage or Fracture	Crystal Habit and Other Notes	Hardness	Name and Chemical Composition
HARD (SCRATCHES GLASS) — CLEAVAGE	Lustre: vitreous Colour: dark green to black Streak: colourless to greenish grey	Cleavage: two planes at nearly right angles	Other Notes: hardness is the same as glass; some samples may scratch glass, others may not	5-6	**PYROXENE** (Augite) X Y (Z$_2$O$_6$), where X = Ca, Na; Y = Mg, Fe, Al; and Z = Si, Al
	Lustre: highly vitreous on cleavage faces Colour: dark green to black Streak: colourless to greenish grey	Cleavage: two planes at angles of 56° and 124°	Other Notes: hardness is the same as glass; some samples may scratch glass, others may not	5-6	**AMPHIBOLE** (Hornblende) X$_{2-3}$Y$_5$Si$_6$Z$_2$O$_{22}$(OH)$_2$ where X = Ca, Na; Y = Mg, Fe, Al; Z = Si, Al
	Lustre: vitreous to pearly Colour: Grey white to blue grey; often exhibits iridescence Streak: white or colourless	Cleavage: two planes at close to right angles	Other Notes: striations common on good cleavage surfaces	6-6.5	**Ca-RICH PLAGIOCLASE FELDSPAR.** (Labradorite) CaAl$_2$Si$_2$O$_8$
HARD (SCRATCHES GLASS) — NO CLEAVAGE	Lustre: adamantine to vitreous on unweathered surfaces Colour: various colours, usually brown, grey, pink, or blue; typically dark, but some samples are lighter Streak: harder than the streak plate, so no streak	Cleavage: none Fracture: breaks easily across crystals	Crystal Habit*: six-sided, barrel-shaped crystals, sometimes with deep horizontal striations	9	**CORUNDUM** Al$_2$O$_3$
	Lustre: vitreous to resinous Colour: varies with composition; most are red to red-brown, but also yellow-green Streak: white to colourless	Cleavage: none Fracture: uneven to conchoidal	Crystal Habit*: commonly occurs in crystal form with 12 or 24 sides	6.5-7.5	**GARNET** X$_3$Y$_2$(SiO$_4$)$_3$ where X = Ca, Mg, Fe, Mn, and Y = Al, Fe, Cr
	Lustre: vitreous Colour: brown to grey black Streak: colourless	Cleavage: none Fracture: strongly conchoidal	Crystal Habit*: prismatic crystals with striations perpendicular to the long dimension	7	**SMOKEY QUARTZ** SiO$_2$
	Lustre: vitreous Colour: olive to greyish green, brown; samples are typically dark-coloured, but some are lighter Streak: pale green to white	Cleavage: none Fracture: conchoidal, but not well expressed in small samples		6.5-7	**OLIVINE** (Mg, Fe)$_2$SiO$_4$
SOFT (DOES NOT SCRATCH GLASS) — CLEAVAGE	Lustre: pearly, to vitreous Colour: dark brown to black Streak: colourless to green/brown	Cleavage: 1-D forming elastic sheets	Crystal Habit*: foliated masses or thin sheets	2.5-3	**BIOTITE** K(Mg,Fe)$_3$AlSi$_3$O$_{10}$(OH)$_2$
	Lustre: vitreous Colour: dark green to black Streak: colourless to greenish grey	Cleavage: two planes at nearly right angles	Other Notes: hardness is the same as glass; some samples may scratch glass, others may not	5-6	**PYROXENE** (Augite) X Y (Z$_2$O$_6$), where X = Ca, Na; Y = Mg, Fe, Al; and Z = Si, Al
	Lustre: highly vitreous on cleavage faces Colour: dark green to black Streak: colourless to greenish grey	Cleavage: two planes at angles of 56° and 124°	Other Notes: hardness is the same as glass; some samples may scratch glass, others may not	5-6	**AMPHIBOLE** (Hornblende) X$_{2-3}$Y$_5$Si$_6$Z$_2$O$_{22}$(OH)$_2$ where X = Ca, Na; Y = Mg, Fe, Al; Z = Si, Al

*Remember that the crystal habit will only be expressed if the sample has grown in unconfined space; most samples will not exhibit this.

Two Questions for You

1. You are presented with a mineral that looks like the one to the right.
    a. Does it have a metallic lustre?
    b. If not, is it light- or dark-coloured?

   Go through the list of questions and try to identify this mineral. It is quite soft, it definitely does not scratch glass. When weak acid is dropped on it, it fizzes.

2. Another mineral is presented to you. It has a dull, silver appearance, but it looks more like an old quarter than a piece of glass. It is very soft; your fingernails can scratch it. It appears to have parallel, wavy lines on its sides, but it's difficult to tell for certain. When you check out its streak it leaves a wide line of very dark grey colour. It feels kind of greasy, and when you put it down you notice that you have black smudges on your hands.

---

1. This sample is non-metallic and light coloured. We are told that it is soft. It fits, therefore into the bottom half of Table 30.3. Careful examination of the photo indicates that the two samples of this mineral express the same angles to their sides. They look like squished cubes; they have 3-D cleavage, but not at 90°, so they are rhombic. So, this mineral fits in the bottom half of Table 30.3 in the section with cleavage. This mineral could be halite, calcite, muscovite, talc or barite. The notes under Calcite, however, say "Cleavage: perfect rhombohedral (3-D, not at 90°) Other Notes: effervesces in cold dilute HCl". This mineral is calcite.

2. The second mineral looks like a dull coin, so it has a metallic lustre. Its streak is very dark grey, so it must be one of the five minerals at the bottom of Table 30.2. Because it's very soft, it must be either Galena or Graphite, with graphite as the softest. It does not have cubic crystals, but it does mark paper and feel greasy. This mineral is graphite.

Igneous Rocks

Igneous rocks crystallize from molten magma or lava in one of the environments of origin described in Table 30.5.

Table 30.5. The relationship between the environment of origin of igneous rocks and their texture.

Environment of Origin	Resulting Texture
Slow crystallization from magma deep within in the earth, which produces large crystals easily seen by the unaided human eye.	**Intrusive, or Phaneritic** (coarse-grained and visible)
More rapid solidification from lava on or near the earth's surface, forming small crystals that are not individually discernable without magnification.	**Extrusive, or Aphanitic** (fine-grained and not-visible without magnification)
Initial, slow crystallization followed by more rapid solidification as partially cooled magma is moved toward Earth's surface, forming two crystal sizes: large **phenocrysts** which formed slowly at depth are surrounded by a matrix of small, infilling crystals which formed more quickly near the surface.	**Porphyritic** (two grain sizes: phenocrysts in a finer matrix)
Fragmental material expelled explosively into the air from a volcanic eruption that falls back to the surface and welds together to greater or lesser extents.	**Pyroclastic** (air-fall deposits of various sizes)
Rock formed very quickly from such rapid chilling of lava at Earth's surface that atoms are 'frozen' in place within the lava, forming glass rather than discernable crystals.	**Glassy** (lustrous and smooth, or frothy like the top of a cappuccino)

You may have heard the phrase 'every rock tells a story.' This is clearly illustrated by the association of each igneous environment of origin with a given texture. You can pick up any igneous rock and know something about its history simply by looking at the grain size. The association is so strong that texture is one of the two criteria by which igneous rocks are classified. The other is chemistry.

Igneous rocks can be divided into three categories based on their chemistry. **Felsic** rocks are rich in the light-coloured feldspar and silica-rich quartz minerals. In contrast, **mafic** rocks are dominated by the dark-coloured *ferromagnesian* minerals rich in magnesium and iron (ferric), such as olivine, pyroxene, and amphibole. If igneous rocks are light in colour we say that they are felsic. If they are dark in colour, they are mafic, and if they contain relatively even mixes of both mafic and felsic minerals, they are said to be **intermediate** in their chemistry.

Combining these two criteria, texture and chemistry, gives an elegant classification scheme for igneous rocks. You may have noticed that Table 30.5 presents a sequence of environment from great depth to Earth's surface. The classification scheme illustrated in the following chart represents these same environments in reverse order, mimicking reality from the surface to depth.

Environment and Texture	Rock Types			
Glassy (almost instantaneous cooling)		Cooled so rapidly that atoms were not arranged into a crystalline structure. These rocks are not classified according to colour because light does not penetrate them well and even felsic glassy rocks appear dark-coloured. **Obsidian**: Glassy rock that displays conchoidal fracture (like a pitted windshield) when broken. **Pumice/Scoria**: Frothy like the top of a cappuccino. Floats on water due to many holes (vesicles) formed from escaping gases. Glassy threads are visible with a microscope. Called pumice if felsic; scoria if mafic.		
Pyroclastic (explosive eruption)	**Ash**: Very fine-grained, loose material less than 2 mm in diameter. **Lapilli**: Pebble-sized, loose material between 2–62 mm diameter. **Tuff**: Ash- or lapilli-sized particles cemented together. **Volcanic Breccia**: Angular pyroclastic fragments cemented in a fine matrix.			
Porphyritic (two grain sizes)	Rhyolite Porphyry	Felsite Porphyry	Andesite Porphyry	Basalt Porphyry
Extrusive/Aphanitic (fine-grained)	Rhyolite	Felsite	Andesite	Basalt
Intrusive/Phaneritic (coarse-grained)	Granite	Granodiorite	Diorite	Gabbro

⟵ Felsic ─────────── Intermediate ─────────── Mafic ⟶

85% felsic minerals   ~ 60% felsic   ~ 40% felsic   ~ 30% felsic

Constituent Minerals

POTASSIUM FELDSPAR, or ORTHOCLASE (Salmon)

QUARTZ (White to Clear)

Ca-rich (blue-grey)

PLAGIOCLASE FELDSPAR

Na-rich (white-grey)

BIOTITE (Black)

AMPHIBOLE (Black)

PYROXENE (Dark Green)

OLIVINE (Green)

If you were given the igneous rock sample pictured to the right, could you give it a name from this classification scheme? Visible crystals of different-coloured minerals are present, so it is coarse-grained and must have cooled slowly, at depth. Therefore, it is an intrusive rock with a phaneritic (visible) texture. It has a 'salt-and-pepper' appearance, which means that it contains a mix of felsic and mafic minerals. It must be in the 'intermediate' chemistry category, then. The felsic minerals are more plentiful than the mafic ones, but there is not a lot of orthoclase feldspar. The closest fit would be granodiorite.

A Question for You

Now it's your turn. What name would you give to the rock on the right?

No visible crystals are present, it does not seem to contain pyroclastic fragments and does not have a glassy texture, therefore, it must be an extrusive rock with an aphanitic (non-visible) texture. The many small holes (vesicles) are caused by gases escaping as lava solidified. They confirm this rock's environment of origin as extrusive. The rock seems quite mafic. A mafic, extrusive rock is basalt.

Sedimentary and Metamorphic Rocks: A Supplement to Unit 31

This supplementary unit is designed to provide additional information for Unit 31 *Physical Geography* that will complement the sedimentary and metamorphic rock lab assignments you may encounter in your physical geography course.

Sedimentary Rocks

Sedimentary rocks are classified on the basis of the type and texture of the particles they contain. They are formed through the:

1. deposition, compaction, and cementation of fragments of pre-existing rocks (inorganic sediment),
2. deposition of organic sediment (fragments of the remains of former organisms), or
3. precipitation of 'evaporites' as water evaporates from a chemical solution.

The type of material they contain, therefore, can be classified into three categories that correspond with the three points made above: inorganic sediment, organic sediment, and chemical precipitates. As pointed out in Unit 31 of the text, sedimentary rocks made from fragments of other rocks are said to be **clastic**, while those from organic or precipitated materials are **non-clastic**.

Study Table 31.1 below and notice how the texture and origins of sediments are combined to give specific names to sedimentary rocks. Note that each source of sediment may consist of various sediment sizes depending on the processes that are in action. For example, high-energy river environments have a tendency to produce coarse-textured sedimentary rocks like conglomerate. Shale will form in low-energy environments where clay settles out of suspension in calm water, such as in an ice-covered lake in winter.

Table 31.1 Classification of common sedimentary rocks based on sediment type and texture.

ORIGIN & TEXTURE	PARTICLE DIAMETER		COMPOSITION AND CHARACTERISTIC FEATURES	ROCK NAME
INORGANIC SEDIMENT (Clastic Texture)	>256mm_	Boulders	Granule to cobble-sized fragments of any type of pre-existing rock cemented by a finer matrix of sedimentary grains.	CONGLOMERATE (if large fragments are rounded)
	64mm_	Cobbles		
	4mm_	Pebbles		BRECCIA (if large fragments are angular)
		Granules		
	½ - 2 mm_	SANDS: Coarse	Sandstones are composed of cemented sand grains. Colours range from white to buff, pink, brown or tan. Different names are given depending on specific proportions of minerals.	QUARTZ SANDSTONE (90% of grains are quartz) ARKOSE (<25% feldspar) GRAYWACKE (15% is fine-grained)
	¼-½ mm_	Medium		
	1/16-¼ mm_	Fine		
	1/256 mm--	"FINES": Silt	Mainly silt-sized particles, which feel gritty to the touch, but are too fine-grained to be seen.	SILTSTONE
		Clay	Mainly clay-sized particles, which feel smooth to the touch. Soft; can be scratched with fingernails. Non-fissile (contains no wavy laminations).	MUDSTONE
			Mostly clay. Is fissile (has wavy laminations) and smooth. May contain fossils.	SHALE
ORGANIC SEDIMENT (Bioclastic Texture)	Very fine		Extremely fine-grained. Composed of $CaCO_3$ remains of microscopic marine organisms. Fizzes with dilute HCl. White to gray.	CHALK
	Variable		Composed of $CaCO_3$ organic remains of shell or coral fragments in a crystalline matrix. Fossils may be visible or not. Effervesces with HCl,	FOSSILLIFEROUS LIMESTONE
INORGANIC CHEMICAL PRECIPITATES (Crystalline Texture)	Medium to Coarse		Chemical precipitate of $CaCO_3$ from seawater. Salty taste. May contain some fossil material (e.g. shell fragments). Effervesces with HCl,	CRYSTALLINE LIMESTONE
	Fine to Coarse		Chemical precipitate of NaCl from evaporation of seawater or groundwater. Extremely salty taste. White to grey colour, but may contain impurities. Cleaves in three directions at 90° angles to form cubes.	HALITE (Rock Salt)
	Fine to Coarse		Common chemical precipitate of $CaSO_4 \cdot 2H_2O$ due to evaporation of sea water. Usually scratches with a fingernail. Variable colour, commonly pink, white or buff.	GYPSUM

Examine the sedimentary rock displayed in Figure 31.1 of the text. The sediments are clearly formed from fragments of other mineral rocks, i.e. this is originates from inorganic sediment. Notice how large the sediments are; they obviously have a gravel or larger texture. Some of the sediments are rounded while others are sub-angular, the latter implying that they did not travel far from their source and have thus preserved their original shape. With the aid of Table 31.1, we can conclude that this sedimentary rock is conglomerate. If the majority of the particles had been strongly angular, this would be a breccia.

Metamorphic Rocks

Two main categories of metamorphic rocks can be recognized on the basis of their structural features. **Foliated** metamorphic rocks contain a parallel alignment of grains. If you examine them carefully, you will observe some linear elements to their structure. They can be subdivided into subgroups:

1. those that are **non-layered**, and:
    a. contain **slaty cleavage**, such that they tend to split along nearly flat, parallel planes like the slate we see used in billiard tables or in flooring or roofing material, or
    b. are **schistose**, with big platy minerals such as mica lined up parallel to one another, and
2. those with **layering**, meaning that they contain coarse-grained, alternating light and dark mineral bands.

Non-foliated metamorphic rocks are massive and structureless, except for some elongated grains. No obvious subgroups are recognized within them.

A simple classification scheme for metamorphic rocks is presented below in Table 31.2. Use it to identify the rock shown in Figure 31.11 of the text.

Table 31.2. A classification scheme for metamorphic rocks.

TEXTURE		COMPOSITION AND DISTINGUISHING FEATURES	ROCK NAME
FOLIATED (contain some linear features)	Non-Layered	Very fine-grained with an earthy luster. Splits easily into flat, parallel sheets.	SLATE
		Very fine-grained with a wavy surface and silky sheen. Tends to split along slightly wavy, parallel surfaces.	PHYLLITE (intermediate form between slaty cleavage and schistosity)
		Contains large, platy minerals in a parallel alignment.	SCHIST
	Layered	Contains alternating bands of light- and dark-coloured minerals that may be more or less folded or contorted.	GNEISS
NON-FOLIATED (contain no linear, or planar, features)		Contains fused quartz grains with a crystalline appearance.	QUARTZITE
		Composed of large CaCO$_3$ crystals with a highly crystalline appearance. Fizzes with dilute HCl.	MARBLE
		A fused conglomerate with stretched, elongated pebbles.	METACONGLOMERATE

The rock shown in Figure 31.11 of the text rock has distinct, parallel bands, so it is foliated. The bands occur in alternating light and dark colours, so therefore, the rock is gneiss (a German word—the g is pronounced, and the e is silent while the i is pronounced, so it is similar to gnu, only 'g-nice' like an Australian 'g'day'). Gneiss work!

Section A: Working with Topographic Maps – Locational Systems

You are moving into the geomorphology section of the course where you will spend lab time examining landforms and landscapes. Because landforms are portrayed on topographic maps, it is essential to have a working knowledge of maps and their symbols. This supplementary unit provides background information, sample calculations and practice questions on:

1. maps of the National Topographic System (NTS) of Canada, and
2. locational systems on Canadian topographic map, specifically, the:
    a. Universal Transverse Mercator (UTM) grid, and
    b. the spherical coordinate system of latitude and longitude.

Maps of the National Topographic System (NTS) of Canada

Canadian topographic maps are organized and labelled according to a grid superimposed over the nation containing numbered **primary quadrangles**. These are subdivided into sixteen smaller units labeled A through P (Fig. A.1). Each of these is further subdivided into smaller segments numbered 1–16 (Fig. A.2).

Canadian NTS map sheets typically are produced at two scales:

- 1:250,000 maps of primary quadrangle subdivisions, e.g. 31C (Fig. A.1), and
- 1:50,000 maps that are cited by combinations of number and letters, e.g. 31C/5 (Fig. A.2).

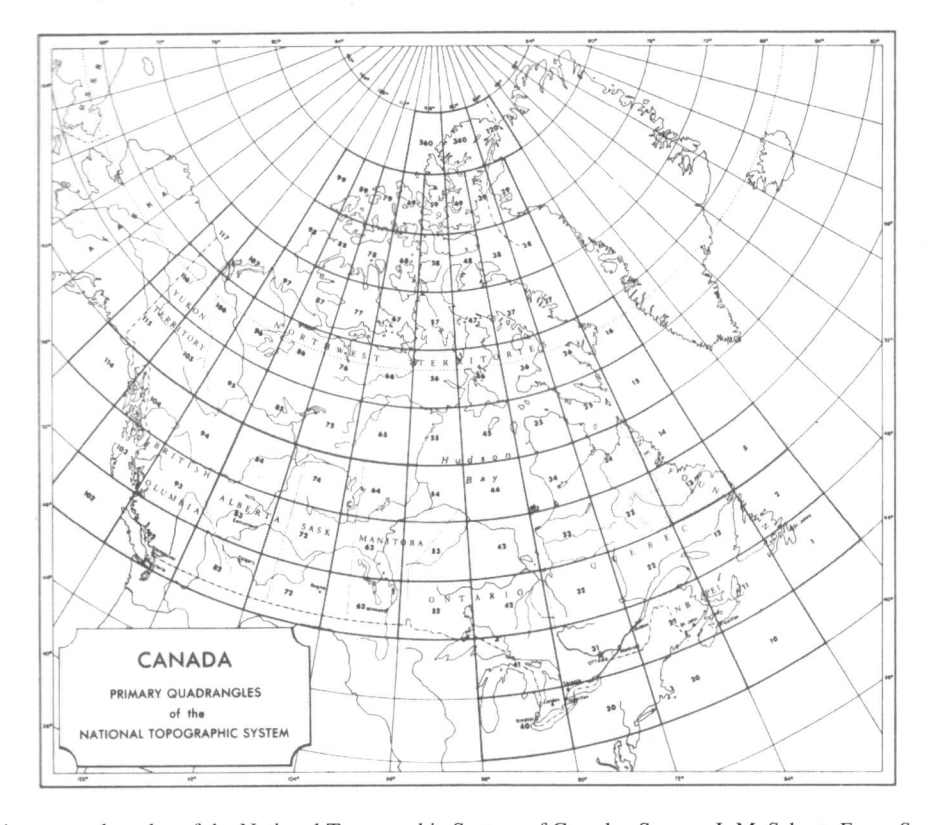

Figure A.1. Primary quadrangles of the National Topographic System of Canada. Source: L.M. Sebert, *Every Square Inch: The Story of Canadian Topographic Mapping*. Department of Energy, Mines and Resources, 1970), Catalogue M52-2974. © 2006. Produced under licence from Her Majesty the Queen in Right of Canada, with permission of Natural Resources Canada.

If your city is located in the shaded portion of Figure A.2, it would be on the 1:50,000 NTS map labelled 62 C/8. The 1:50,000 NTS map directly to the west of it would be 62 C/7. Because the subdivisions shown in section C are repeated through every subdivision of every primary quadrangle, the map to the east of it would be 62 B/5.

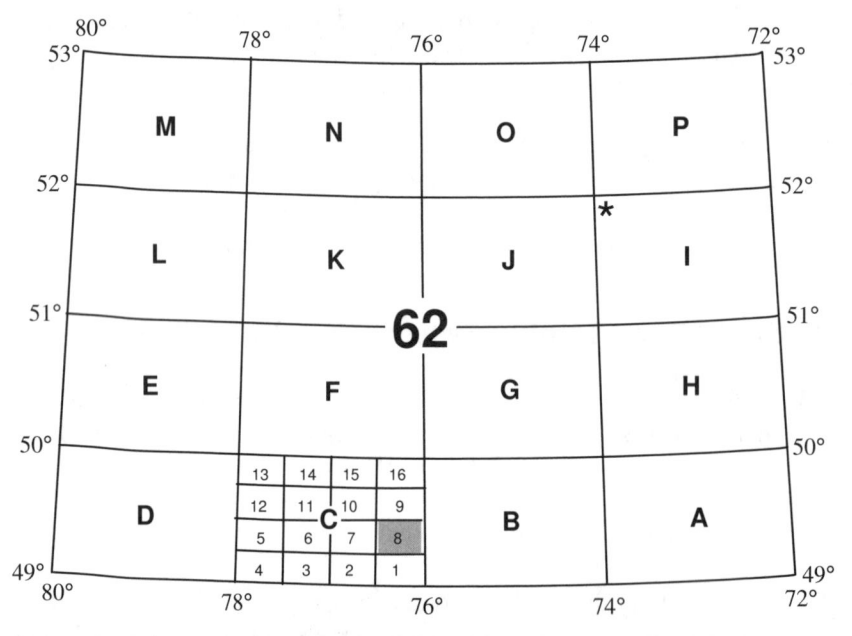

Figure A.2. Primary quadrangles are subdivided into sixteen smaller units labeled A through P, each of which is further subdivided into 16 smaller segments numbered 1–16. Maps of the 1:50,000 series are labeled with number, letter, number codes such as 62 C/8.

NTS map sheets follow particular colour conventions:

- Hydrographic (water) features are always shown in blue. The names of water bodies are italicized.
- Relief features (elevation) are shown in brown.
- Vegetation is coded with the colour green.
- Cultural features (e.g. roads, towns, administrative boundaries, etc.) are shown in black and red.

Some Questions for You

1. If your home was located at the asterisk in Figure A.2, what 1:250,000 map sheet would it be on?
2. If your home was located at that asterisk, what 1:50,000 map sheet would it be on?
3. What 1:50,000 map sheet would be *directly north* of the map your home is on?

Answers are provided at the end of this unit.

The Universal Transverse Mercator (UTM) Grid

Examine any Canadian topographic map (Fig. A.3) and you will see a grid of pale blue lines used to communicate locations. In the northern hemisphere, this UTM grid is superimposed on each zone such that the origin of the grid is the equator and the western edge of map zone (see Unit 3 Fig. 3.1). Vertical grid lines that run N–S are called **eastings** because they specify distance east of the grid origin.

Horizontal grid lines running E–W are called **northings** because they indicate distance north of the grid origin. Eastings and northings can be identified on maps (even though they are not labelled with those names) by noting the direction over which they increase in value. In Figure A.3, for instance, the eastings increase to the east (line 65, 66, 67) while the northings increase to the north (lines 84, 85, 86, 87).

Working with the Civilian UTM Grid

Civilian citations of the UTM grid begin with the map's zone number, then continue with the full easting reference in metres east of the grid origin, and the full northing reference in metres north of the grid origin.

In the example map to the right, the feature of interest is the box at approximately easting line 67 and northing line 84. To give the full citation, look in the right margin of the map for the zone number and in the corners of the map for the complete easting and northing coordinates. In this case the zone number is 10. The easting coordinate in full is 467000 metres east of the zone origin, or 467000m.E. Clues to the full easting citation are given in the map margin.

The grid line immediately south of the feature is 54840000 metres north of the zone origin and the feature itself is approximately 1/10th of a grid line past it. Since these are 1000-metre grid squares, it is a further 1/10(1000m) = 100m to the north. The northing coordinate, then, is 5484100m.N.

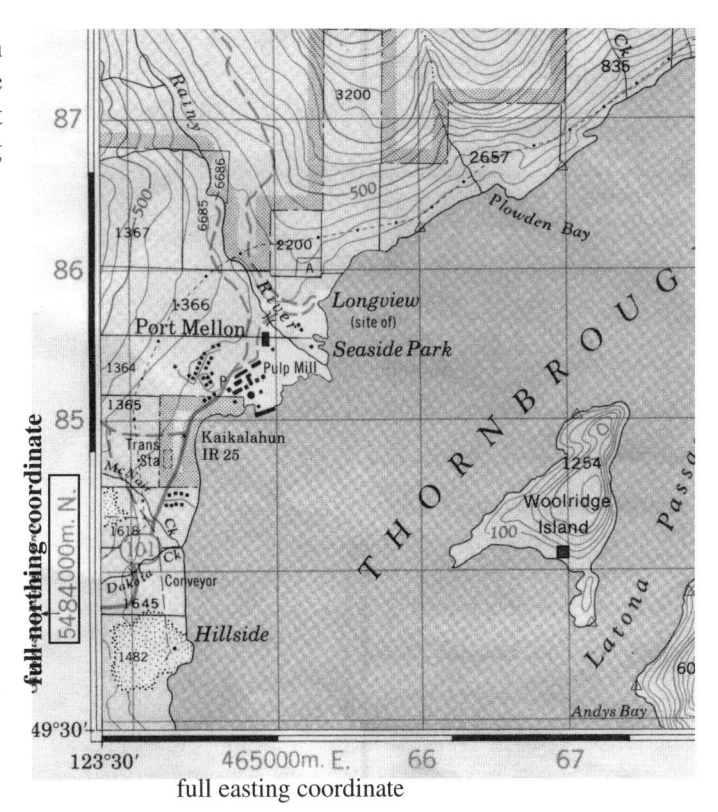

Figure A.3 Easting and northing values on 1:50,000 maps.

The Civilian UTM Grid citation would be Zone 10, 467 000, 5 484 100. In British Columbia, under the Resource Inventory Committee system, the units would also be given, making the complete UTM grid reference for this feature Zone 10, 467000m.E., 5484100m.N.

Working with the Military UTM Grid

The abbreviated military use of the UTM grid differs on 1:50,000 scale maps and 1:250,000 maps. Information on the use of the military UTM grid is presented below for both scales.

Large scale map example

On large-scale maps (1:50,000 and larger) each grid line is identified with a two-digit number (Fig. A.4a). At a scale of 1:50,000 each grid square measures 1000 m x 1000 m, for an area of 1 km^2. To communicate the location of any feature on the map, a 6-digit UTM grid reference is cited: the first 3 digits are easting coordinates, the second 3 digits are northings. As an example, we'll find the military UTM Grid reference for the school in Figure A.4.

Instructions for Citing the Easting Value:
1. The school is located between eastings 56 and 57 (Fig. A.4a). Select the easting that undershoots the location (do not overshoot). In other words, determine the number of the grid line to the west of the feature (Fig. A.4a). This is easting 56.
2. Place a ruler on the map (Fig. A.4b). Note that the distance between UTM grid lines on 1:50,000 maps is exactly 20 mm, which represents a ground distance of 1000 m. Subdivide the distance into tenths. Each tenth will measure 2 mm on the map and will represent a ground distance of 100 m.
3. The middle of the school is exactly 12 mm or 6/10ths past easting 56. The easting grid reference, therefore, is 566. The first two digits indicate the easting grid line while the third digit reveals how many tenths past that line the feature is found.

Instructions for Citing the Northing Value:
1. The school is situated between northings 32 and 33.
2. A ruler shows that the middle of the school building is located nearly 6 mm or 3/10ths past northing 32.
3. The northing grid reference is approximately 323.

The military UTM grid citation for the location of the school is 566323. The first three figures present the easting coordinates while the final three give the northing values. The easting is always given before the northing.

This system provides a quick way of communicating rough locations. It can be fine-tuned by further subdivision of the grid distances into hundredths. For instance, the centre of the church building is exactly 6/10ths beyond easting 56, but it is a little beyond 3/10ths past northing 32. It is at northing 32 + 3/10ths + 8/100ths. Its location could be cited as 56603238. The first four figures are eastings values; the last four are northing measures. This 8-figure UTM grid citation adds greater precision and aids in pinpointing the location of small features on maps.

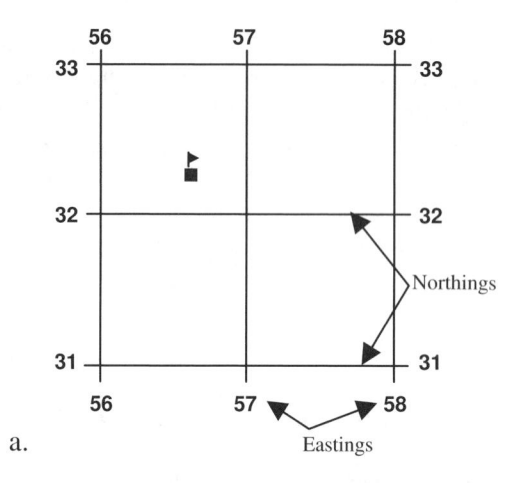

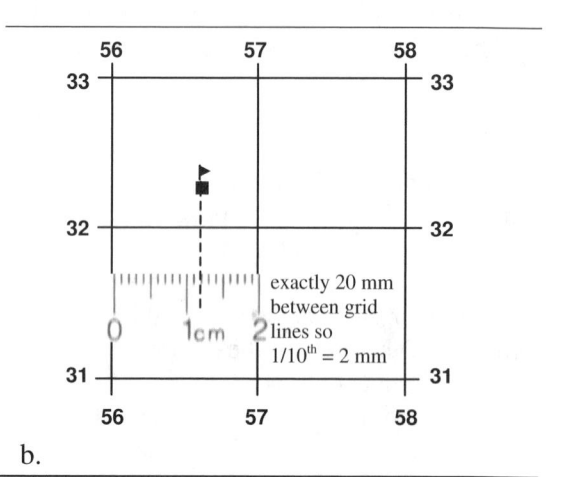

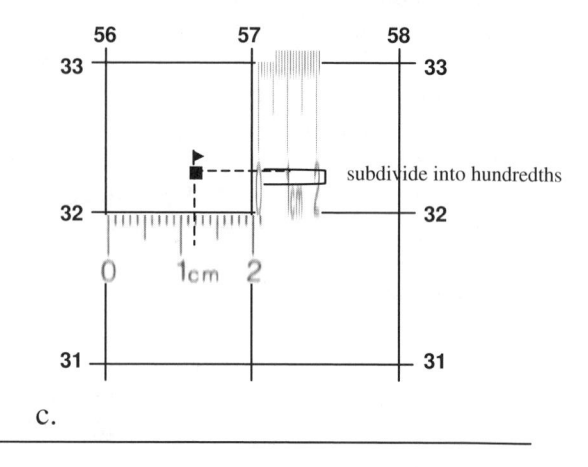

Figure A.4 The eastings and northings of the UTM Grid on 1:50,000 scale maps. The location of specific features can be pinpointed by subdividing the distance between grid lines (1000 m) into tenths.

Small scale map example

On medium and small scale maps (1:100,000 and smaller) each grid line is identified with a one-digit number (Fig. A.5). These abbreviated grid numbers repeat themselves every 100,000 metres, so specific 100,000 metre grid segments on the map are identified with letters. A 1:250,000 scale military UTM citation will always begin with two identifying letters. In the example given in Figure A.5, we can use a ruler to measure how many tenths past easting line 9 the school lies (notice that grid lines are 40 mm apart). We can do the same to identify the northing value. The grid reference for the school in Figure A.8 would be NK9396.

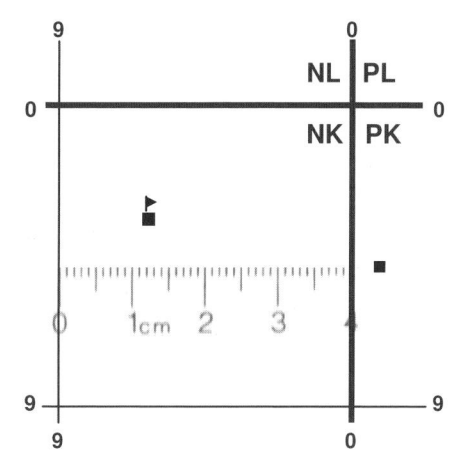

Figure A.5. The UTM or military grid at a scale of 1:250,000.

Some Questions for You to Try

1. Examine Figure A.3. What are the civilian UTM coordinates for the stream as it enters Plowden Bay?

2. What are the 6-digit military UTM coordinates for the stream as it enters Plowden Bay?

3. Examine Figure A.5. What are the coordinates of the rectangular building located on the right side of the figure?

Answers are at the back of the unit.

Web activity

Check out Natural Resources Canada's Topo101 website at http://maps.nrcan.gc.ca/topo101/index_e.php

The Spherical Coordinate System – Latitude and Longitude

Latitude and longitude were examined briefly in Unit 3. Here they are revisited to explore how they communicate location on Canadian topographic maps.

Latitude coordinates tell us (in terms of angular distance) how far north or south a location is from the equator. All Canadian maps are north of the equator. Parallels and their subdivisions are shown as black and white vertical line segments on the sides of the map, which represent one minute of latitude. They increase toward the north, indicating that the maps are in the northern hemisphere. Every fifth minute of arc is labelled.

Longitude coordinates tell us how far east or west a location is from the prime meridian. Meridians and their subdivisions are shown as black and white horizontal line segments at the top and bottom of the maps. Each line segment represents one minute of longitude. All Canadian maps are west of the prime meridian, as indicated by longitude values increasing to the west. Again, every fifth minute is labelled.

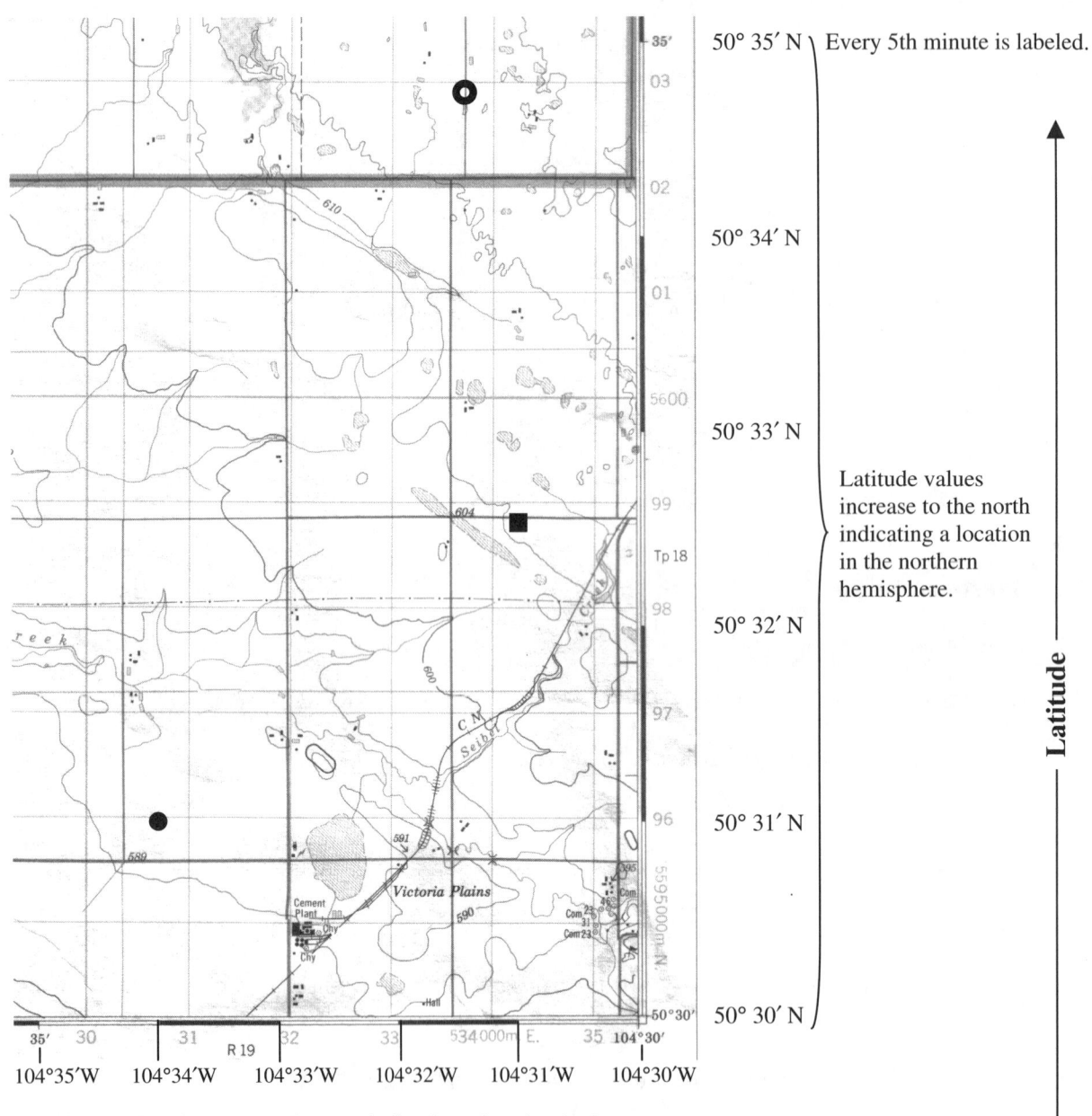

Figure A.6. Part of 1:50,000 NTS sheet of Lumsden, Saskatchewan (72 I/10) showing how degrees and minutes of latitude and longitude are shown on Canadian maps.

The location of the SE corner of the map is 50°30′N, 104°30′W. Remember from Unit 3 that latitude is always given first. The black rectangle is not located in as convenient a spot for citing latitude and longitude coordinates. It is part way between 50°32′N and 50°33′N, so we will have to go to seconds of arc.

You may remember that there are 60 seconds in one minute of arc. Therefore, the location of the black rectangle is approximately 50°32′30″N, 104°31′00″W. More precise values could be given by stretching a ruler all the way across the map at 50°32′N. When you work with maps in your laboratory classes you will notice that the blue grid does not line up precisely with latitude and longitude coordinates, necessitating careful ruler placement when working with latitude and longitude citations. This phenomenon will be explained in Section B.

Some More Questions!

1. What is the latitude/longitude citation for the black circle in Figure 6?

2. What is the latitude/longitude citation for the open circle in Figure 6?

Answers are below.

Answers to the Questions on Primary Quadrangles and Their Subdivisions

1. If your home was located at the asterisk in Figure A.2, what 1:250,000 map sheet would it be on?
   Your home would be on the 1:250,000 map 62I

2. If your home was located at that asterisk, what 1:50,000 map sheet would it be on?
   Your home would be on the 1:50,000 map 62I/13

3. What 1:50,000 map sheet would be *directly north* of the map your home is on?
   The 1:50,000 map directly north of the one your home is on would be 62P/4

Answers to the UTM Grid Coordinates Questions

1. Examine Figure A.3. What are the civilian UTM coordinates for the stream as it enters Plowden Bay?
   Zone10 466400 m.E. 5486500 m.N.

2. What are the 6-digit military UTM coordinates for the stream as it enters Plowden Bay? *664865*

3. Examine Figure A.5. What are the coordinates of the rectangular building located on the right side of the figure?
   PK0195

Answers to the Latitude/Longitude Coordinates Questions

1. What is the latitude/longitude citation for the black circle in Figure A.6? *Roughly 50°31′N, 104°34′W*

2. What is the latitude/longitude citation for the open circle in Figure A.6? *Roughly 50°34′40″N, 104°31′30″W*

Section B: Working with Topographic Maps – Direction Indicators

Continuing an exploration of topographic maps, this supplementary unit provides background information, sample calculations, and practice questions on direction indicators used on Canadian NTS sheets.

Direction Indicators

By convention most maps are arranged with N at the top. But it's not that simple. All Canadian topographic maps, and most others produced internationally, show three 'norths' in a **declination diagram** found in the map margin. The 3 north indicators refer to:

1. **True North,** which is the north pole of Earth's axis of rotation. The true north indicator of the declination diagram is topped with a star—the North Star.
2. **Grid North**, which refers to the N–S running lines (eastings) of the UTM grid. Remember that the origin of the UTM grid is at the equator. By the time we reach the latitudes of Canada, a small divergence has occurred between grid north to true north. On a declination diagram grid north is topped by a small square. The angle between true and grid north is recorded on the diagram (0°16' in this case).
3. **Magnetic North**, which refers to the direction a compass needle will face in response to the Earth's magnetic field. It is marked by an arrow in the declination diagram. Because Earth's magnetic field is ever changing, map users must calculate the angle to magnetic north at their time of use.

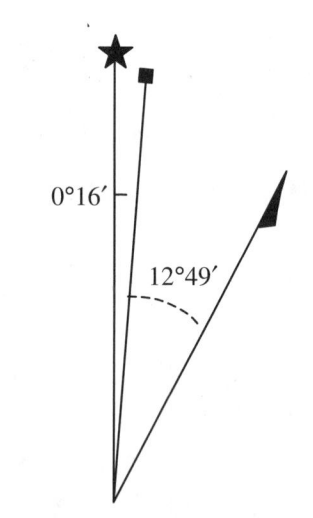

APPROXIMATE MEAN DECLINATION 1999
Annual change decreasing 14.2'

Figure B.1 Fictional declination diagram typically placed in the right margin of Canadian topographic maps.

The angle between magnetic north and true north is known as the **magnetic declination**. Magnetic declination is specified for the centre of Canadian map sheets. It varies temporally, as well as spatially from one map to another. When working with a given map, it is necessary to determine the current magnetic declination.

Calculation of Current Magnetic Declination

Annual magnetic changes are printed with the declination diagram. To calculate the current value, the total magnetic variation (the annual magnetic change multiplied by the number of years since publication) must be added to or subtracted from the declination angle.

Figure B.1 can provide an example calculation. The total declination between magnetic and true north for 1999 was 12°49' + 0°16' = 13°05'. Seven years have passed since the publication of this data (2006–1999 = 7) and the declination has decreased by 14.2' each year. (Note that this angle is expressed in decimals of minutes of arc rather than in minutes and seconds. You will see both systems of expression in the field of geography. Be adaptive.)

The declination for 2006 is:

$13°05' - (7y)(\frac{14.2'}{1y}) = 13°05' - 99.4'$, but there are only 60' in 1°, so 99.4' can be re-expressed.

$13°05' - 99.4' = 13°05' - 1°39.4' = 11°25.6'$ which, if expressed in seconds of arc, would be $11°25'36''$.

Expressing a final answer as either 11°25.6′ OR as 11°25′36″ would be correct. Note in class whether you are required to give a particular format.

If you are performing some geographic fieldwork, you may want to set your compass to line up with the UTM map grid. It would be necessary, in that case, to calculate the difference between magnetic and grid north. You could apply the same principles to that problem.

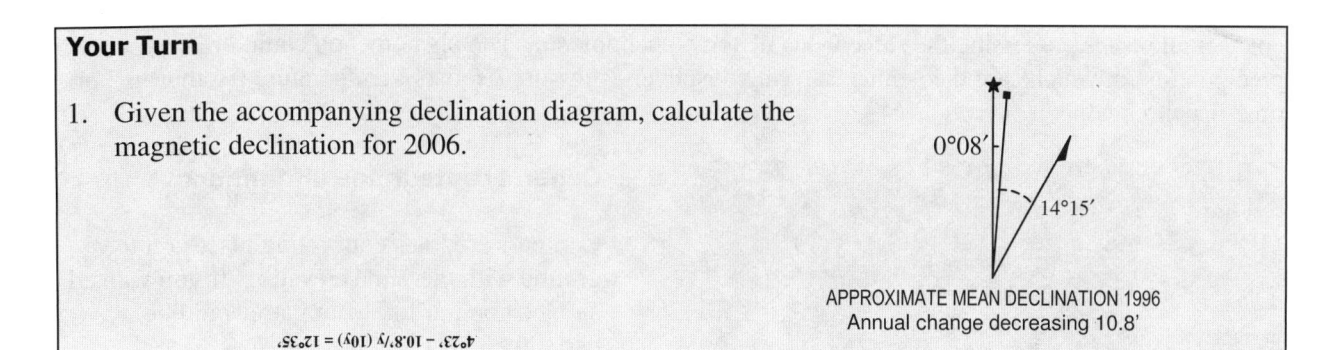

Your Turn

1. Given the accompanying declination diagram, calculate the magnetic declination for 2006.

   0°08′

   14°15′

   APPROXIMATE MEAN DECLINATION 1996
   Annual change decreasing 10.8′

   4°23′ − 10.8′/y (10y) = 12°35′

89

Section C: Working with Topographic Maps – Contour Lines

It is essential that physical geographers be well versed in the symbols of elevation: **contour lines**. Continuing an exploration of topographic maps, this supplementary unit provides background information, sample calculations, and practice questions on contours, specifically on:

1. general contour line information
2. construction of contour lines, and
3. interpretation of contour patterns.

Contour Lines: Citing Elevation on Topographic Maps

Contour lines are a type of **isoline**. As described in Unit 3 of the text, isolines are lines drawn on maps that connect all places possessing the same value of some phenomenon. The phenomenon could be precipitation, in which case the isolines are called isohyets, pressure (isobars), temperature (isotherms), or time (isochrones).

General contour line information

Contours are lines connecting places on the ground with the same elevation. If you walked along a contour line your path would be perfectly level.

Figure C.1 illustrates some fundamental contour line patterns.

- Contours tend to parallel adjacent neighbours.

- Widely spaced contours indicate little change in elevation (flat terrain), for example the coastal plain and the elevated terrace just above the coast in Fig. C.1A and C.1B are relatively flat.

- Closely spaced contours indicate a large change in elevation over a short distance (steep slope, or a cliff if contours almost touch). The cliff marked **X** is an example of steep terrain shown with closely set contour lines.

- Contours make a V-shape around rivers with the V 'pointing' upstream to higher elevation.

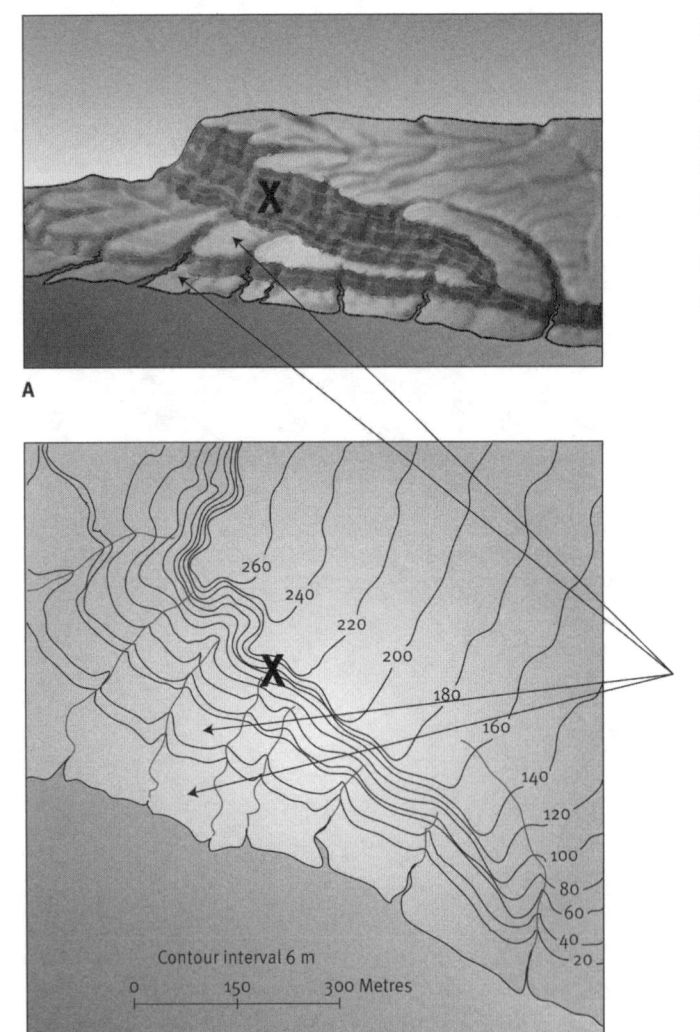

Figure C.1 illustrate important contour patterns (Figure 3.13 in H. de Blij, *Physical Geography*, 2005).

On Canadian topographic maps:

- Contours are drawn as thin, solid brown lines. (On a glacier surface they are dashed.)

- To minimize visual clutter, every 5th line is thickened and labelled with an elevation value. These are called **index contours**. The other lines are said to be intermediate contours.

- Contours are labelled with elevation values in a small break in the lines. The elevation values are oriented such that they 'read uphill', in other words, the top of the numbers face higher ground on the map. This small touch provides readers with an instantaneous sense of higher and lower elevations at a quick glance.

- Contours are plotted on a given map at a consistent interval, e.g. 200, 210, 220, 230 metres, that by extension would pass through zero. The **contour interval** is the elevation difference between adjacent contours, in this example, 10 metres. On Canadian topographic maps, the contour interval is always stated in the bottom margin of the map.

- **Auxiliary contours** are sometimes added to flat terrain where index and intermediate contours would show little detail. By convention auxiliary contours are shown as dashed lines at half the contour interval of the rest of the map.

- Contour lines never cross. That said, at steep cliffs adjacent contours may almost touch.

- Although not necessarily on a single map sheet, each contour will form a continuous (and sometimes highly convoluted) loop. Each contour closes back on itself; no contour simply stops.

Construction of Contour Lines

Geography students are often introduced to contour line construction with a map of spot heights like that given in Figure C.2. Complete this figure by adding contour lines at 50 metre intervals. A hint has been provided in the lower right corner. Begin the completion of the 150 metre line by deciding where it fits between the 140 m and the 180 m spot heights.

Mentally divide the distance between adjacent spot heights to see where each line must go. Continue that process across the map one contour line and one small step at a time. Don't try to see the whole picture at once—just look at the values immediately surrounding your pencil and let the elevation values guide the line placement. Keep an eraser handy!

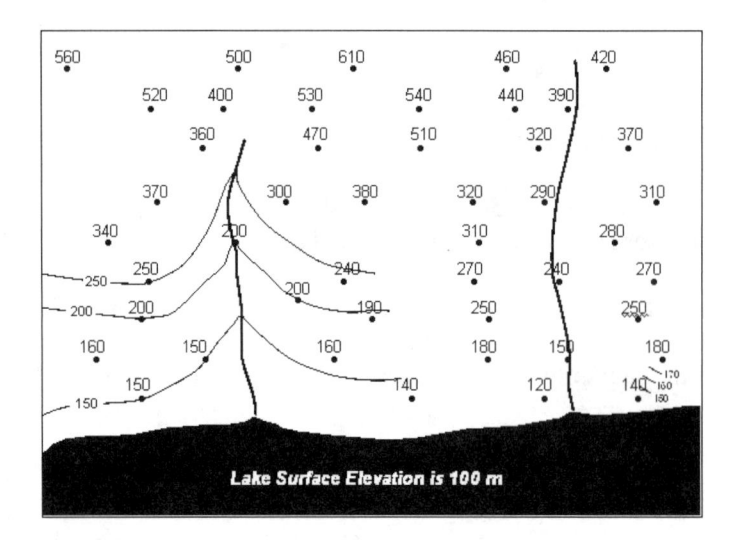

Figure C.2 A partially completed contour map. Continue to add contours at appropriate places among the spot heights to complete the map. Hints have been provided near the bottom right corner to help you begin to place the 150 m contour line.

When you have become more accomplished at drawing contours, you may not be given explicit instructions about which lines to begin contouring with. In that case, first assess the range of values shown on the map. Select a contour interval that will add enough lines to provide detail, but not so many as to be cluttered. The lowest contour must have a value greater than the minimum elevation on the map, and the highest contour must be less than or equal to the highest elevation on the map.

Interpretation of Contour Lines

If you've completed the contour map in Figure C.2, examine the completed version of the figure at the back of this unit. What do you notice about contour line patterns? First, you may recognize that, as in Fig. C.1 examined earlier, these contours tend to parallel their neighbours. Secondly, you may observe that on this map, as in Fig. C.1, contours make a V-shape around rivers, with the V pointing to higher elevations. These are just two of the patterns that a trained eye can pick out. We'll examine some classic contour line configurations in the following section.

Figure C.3 has been reproduced below to help illustrate the relationship between recognizable contour patterns and the landforms they represent. Segments of this figure have been captured and repeated in Table C.1 to help you visualize contour-landform connections.

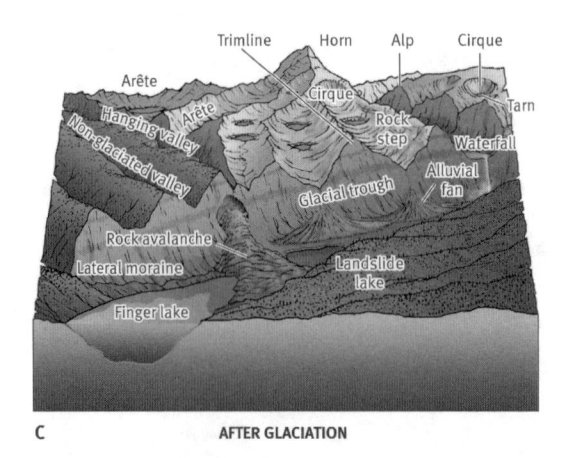

Figure C.3 illustrates some glacial landforms as you would see them in the field (Fig 47.8 in H. de Blij, *Physical Geography*, 2005

Examine Table C.1. Try to visualize the landforms as you would see them on the ground (first column), then look at a contour pattern that represents them (column 2). Try to pick out the connection. Verify your ideas by looking at column 3 for a highlighted version of the contour pattern.

Table C.1. Descriptions of landforms and the contour patterns that represent them on topographic maps.

Contour Pattern and Landform Description	Contour Pattern	Highlighted Pattern
Triangular or pyramidal contours form a **Horn**. A horn is a pyramidal peak associated with alpine glaciers.		
Contours that make a sharp V with the V pointing to lower elevations indicate an **Arete**. An arête is a knife-edged ridge associated with alpine glaciers.		
Contours that make a sharp V with the V pointing to higher elevations show a **River Valley**. This contour pattern represents a V-shaped valley carved by a stream.		
Contours form a truncated U pattern. The U 'opens' toward lower elevations, and its 'base' is backed against higher elevations - **Cirque**. A cirque is a basin-shaped depression carved by alpine glaciers.		

93

Contour Pattern and Landform Description:	Contour Pattern:	Highlighted Pattern:
Contours form an elongated U pattern. The U opens toward lower elevations, and its base can be traced back to a cirque. Contours are widely spaced in the centre of the elongated U and tightly spaced along the outer 'walls' – **Glacial Trough**  A glacial trough is a large U-shaped valley carved by an alpine glacier.		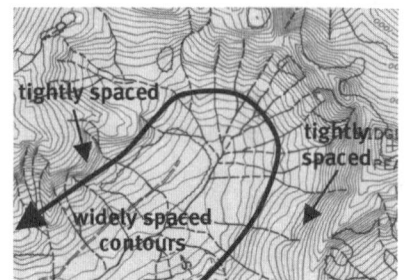
Contours form a U pattern. The U opens toward higher elevations, and its base faces lower elevations - **Non-glaciated Ridge**  This ridge has not been subjected to glaciation and has not been carved into a steep, 'knife-edged' shape.	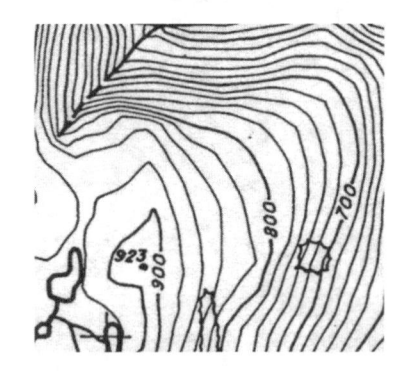	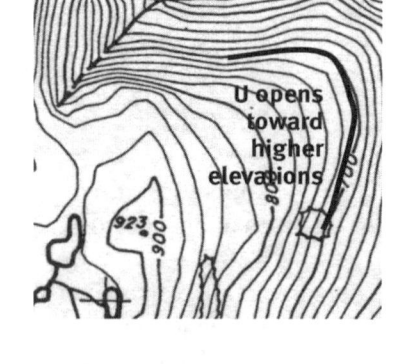
In more general terms: - Closed contours with 'hatching' on the inside of the lines indicate **depressions**. - Closed contours with the highest elevation in the middle indicate a **hill**.	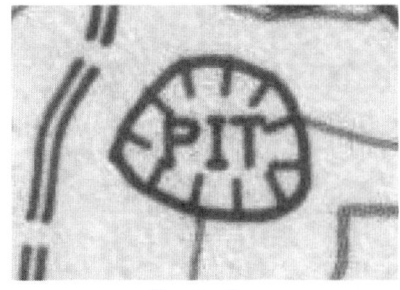 Depression	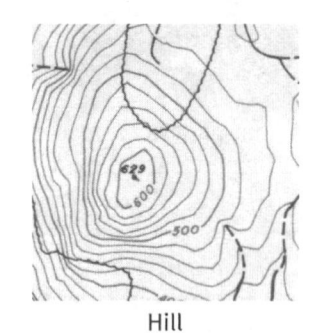 Hill

Some Contour Interpretation Questions for You

Examine the following contour map and outline the following features:
1. arête
2. horn
3. cirque
4. glacial trough
5. small stream valley

3800
Mt
Varley
4100
1000
McNab
T L 13119
2000
1000
4500
3200
Creek
T L 7564
2700
3000
4000
Box
Canyon
Ck
4300
T L 7563
3700
2500
T L
10566
4200
3900
677B
2000
2500

Contour interpretation 'answers'

Here are some potential 'answers' to the question. Note that several of each of these features are shown on this map segment.

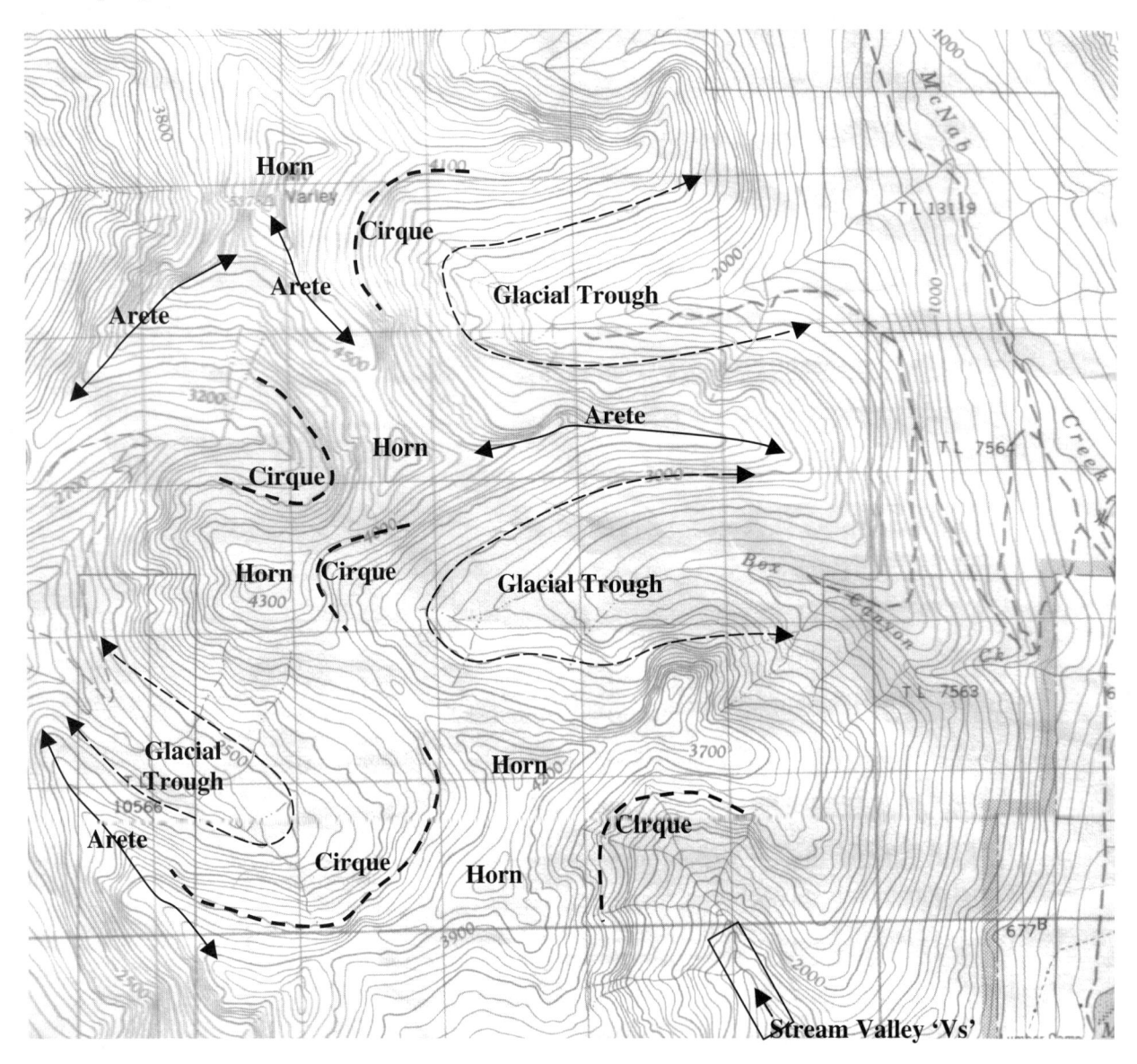

The completed contour map for Figure C.2 is shown below.

How did you do? Your lines may be in slightly different places, but the pattern should be very similar. The highest contour line you should have is 600 m.

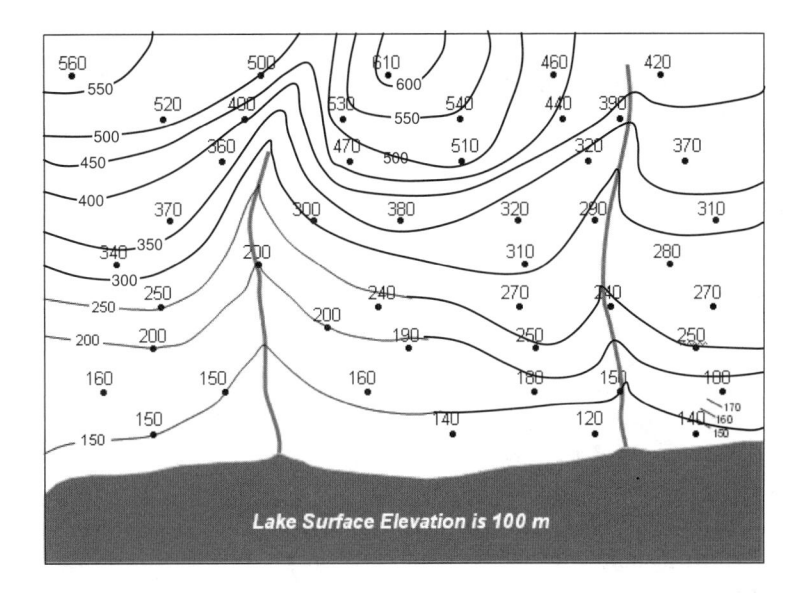

Section D: Working with Topographic Maps—Contour Calculations

Physical geographers are always measuring attributes of landforms shown on topographic maps. This section of the supplement is designed to teach you how to:

1. interpolate elevations between contour lines
2. draw topographic profiles
3. determine the vertical exaggeration of a profile
4. calculate average gradients between two points on a map.

Interpolating Elevations Between Contour Lines

Sometimes you may need to estimate the elevation of a point on a map that does not lie directly on a contour line. This is a process of **interpolation**, where values are estimated *within* a set of known values. In this case, we need to estimate the elevation of the land between two known contour lines. Point X in Figure D.1, for instance, lies between two contours. Careful investigation of the contour values reveals that the contour interval for this map is 20 m. Point X lies more than one contour line below the 500 m contour. The elevation of X, therefore, is less than 480 m, but more than 460 m. That's all we can say with certainty. Because X is halfway between 460 m and 480 m, we can estimate its elevation as 470.

Point Y presents a different scenario. Again, it lies between 460 m and 480 m, but it is not halfway between those neighbouring contour lines. *Listen for guidance from your professor or instructor about how he or she would like you to estimate contour values.* Some people will say that Point Y is approximately ¼ past the 460 m contour line and so its elevation should be cited as 465 m. Others will suggest that anything between contours be taken as halfway between them in elevation.

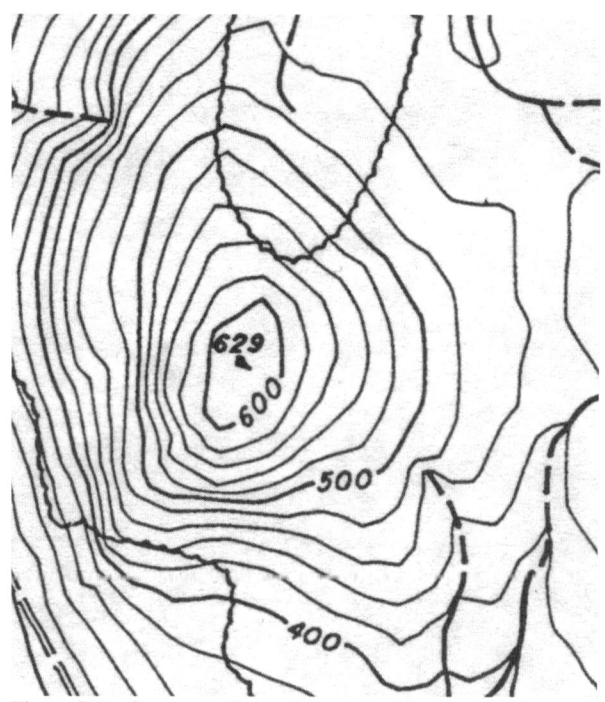

Figure D.1 Segment of a contour map with a contour interval of 20 metres.

Drawing a Topographic Profile

Topographic profiles show cross-sections of the land surface, as you would see them in silhouette if you were on the ground with an unimpeded view. You will likely be asked to create one in the geomorphology section of a first year physical geography program. Instructions, along with some graphical communication, for drawing a topographic profile are presented below. Figure D.2 will be used as an example topographic map from which the profile is to be drawn. The profile is to be taken along the line between Points X and Y.

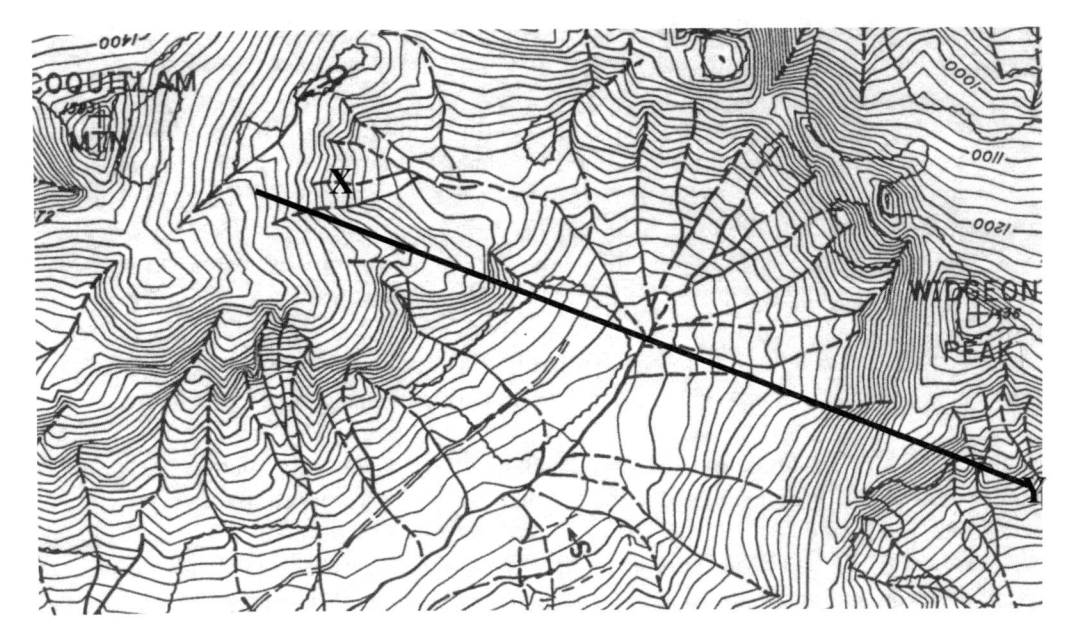

Figure D.2 Topographic map from which a topographic profile will be constructed between points X and Y.

To construct a topographic profile, perform the following steps:

1. Place an unused piece of paper along the line between A and B. You might want to lightly tape it down with masking tape. You could also fold the paper to set a firm edge against the X-Y line.

2. Clearly mark the intersection of the two end points (X and Y) with your paper. If your paper should move, this will allow you to reposition it precisely.

3. Mark the intersection of *every* contour line with your paper (Figure D.3). The only exception to this rule will occur if the line X-Y intersects with extremely steep terrain. In that situation only, it is permissible to mark off just the index contours (every 5th, thickened line).

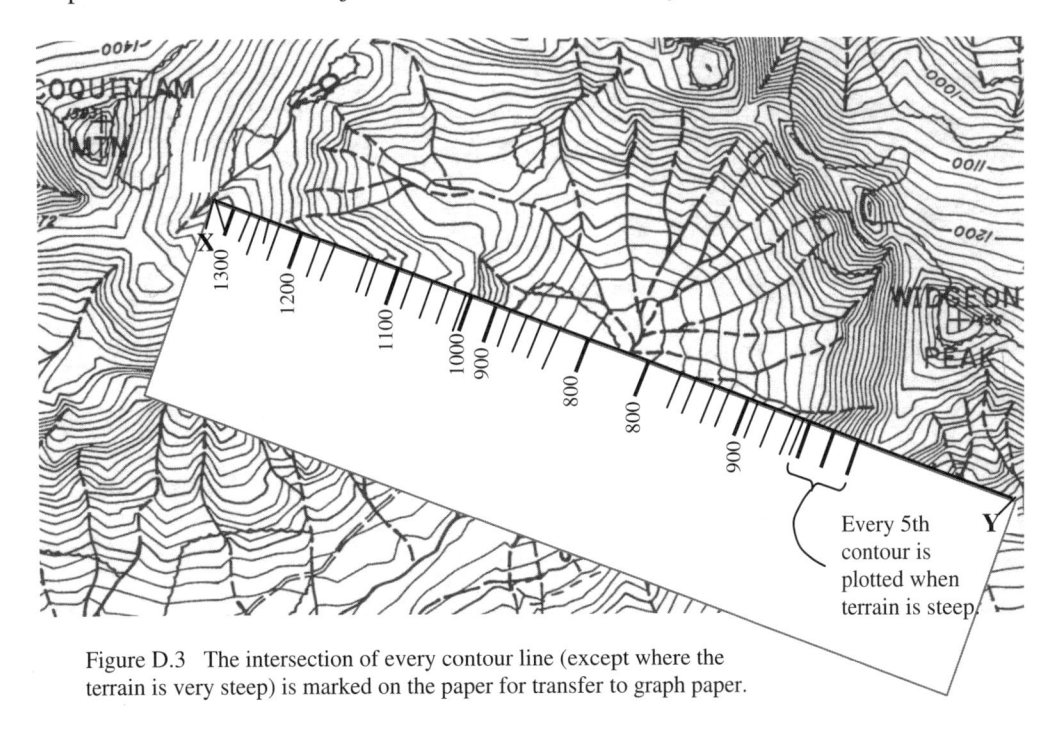

Figure D.3 The intersection of every contour line (except where the terrain is very steep) is marked on the paper for transfer to graph paper.

4. Note the elevation of each index contour on your paper. Because you know the contour interval you can then determine the elevation of each line. Watch out for contour lines that loop back!

5. Thinking through the hints for preparing quality graphs outlined in Unit 4 of the supplement, begin to set up a graph on which you will plot the profile data. The horizontal axis represents ground distance and takes its scale from the map. It must be the same length as line X-Y. The vertical scale represents elevation. You may be given explicit scale instructions for the vertical scale. If so, use them. If not, be prepared to go through a trial and error process of mocking up a profile, then doing it again with an adjusted vertical scale. Ideally, you want to use a vertical scale that provides a realistic silhouette of the landscape. You do not want to add so much vertical stretching that you make mountains out of small hills. Nor do you want to create a squished, flat line of a profile that provides no detail of elevation change.

6. Place your paper along the horizontal axis of your graph and plot the elevation data with a series of small points (Figure D.4). When finished, connect the points with a smooth line that provides a realistic silhouette of the landscape. It may contain small dips for stream channels, for instance. Two points in a row at the same elevation indicate that the land between them probably rises or falls slightly. After looking at the trend in the landscape on either side of those points, you have to decide whether the elevation falls (indicating a valley bottom) or rises (signalling the crest of a hill).

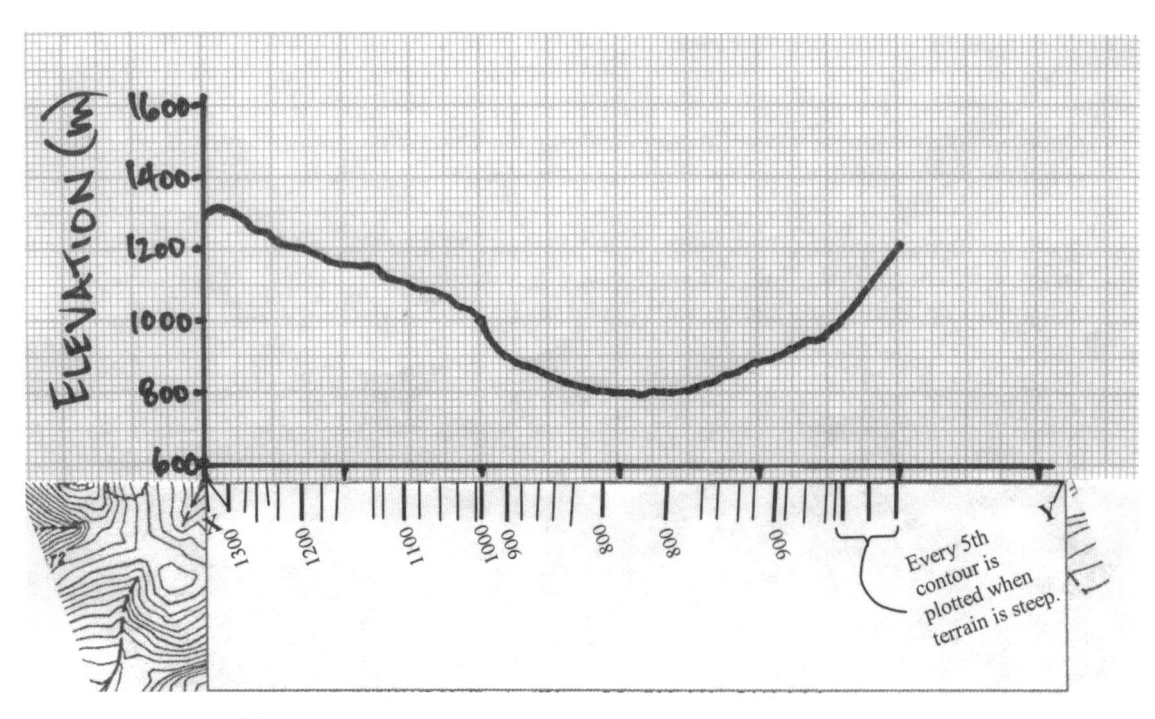

Figure D.4 Plot the elevation data on the graph to construct the topographic profile.

7. Give the profile an informative title including the general location, map sheet and UTM grid coordinates that mark the boundaries of the profile (points X and Y).

8. If possible, indicate the orientation of the profile with a north arrow.

9. Cite the vertical exaggeration of the profile. (Instructions are given below).

Calculation of Vertical Exaggeration

Let's assume that you have completed the topographic profile. You used a vertical scale of 1 cm represents 200 m. The scale of the map the profile is extracted from is 1:20,000. At this scale, 1 cm represents 20,000 cm, or 200 m. So the scales are identical. Therefore, the vertical exaggeration is 1.

The vertical exaggeration communicates the number of times the vertical scale is larger than the horizontal scale. In relatively flat terrain it would be important to put in some vertical exaggeration to see any elevation change. Because the creator of the profile has choice about whether and how much vertical exaggeration to include, the vertical exaggeration is always cited on a topographic profile to give readers a sense of how much vertical stretching has been involved.

To calculate the vertical exaggeration, the distance 1 cm represents on the horizontal axis is compared to the distance 1 cm represents on the vertical axis. For example, if 1 cm represents 500 m on the horizontal scale and 200m on the vertical scale, the vertical exaggeration can be calculated as:

$$VE = \frac{\text{Horizontal Scale}}{\text{Vertical Scale}} = \frac{500m}{200m} = 2.5$$

In this example the vertical scale is stretched 2.5 times more than the horizontal scale.

A Question for You

1. You are creating a topographic profile from a map with a scale of 1:40,000. You use a vertical scale such that 1 cm represents 100 m. What is the vertical exaggeration of your profile?

   1 cm on the horizontal scale represents 40,000 cm or 400 m.
   1 cm on the vertical scale represents 100 m.
   VE = 400 m / 100 m = 4 times

Calculation of Gradient

The gradient refers to the steepness of a slope (remember rise over run?). It can be expressed in a variety of ways: as a ratio, percentage, fraction or degree. The gradient is defined as the:

$$Gradient = \frac{\text{Change in Elevation } (\Delta E)}{\text{Change in Horizontal Distance } (\Delta D)}$$

We'll work through the same example four times to illustrate different expressions of gradient. In each case we'll assume that the change in elevation is 200 m while the change in horizontal distance is 800 m.

Gradient as a ratio

When the gradient is expressed as a ratio it represents the change in horizontal distance associated with a vertical change of one unit. We want a final answer expressed as a ratio of 1:X.

$$Gradient = \frac{\text{Change in Elevation }(\Delta E)}{\text{Change in Horizontal Distance }(\Delta D)} = \frac{200m}{800m} = \frac{1}{4}$$

The gradient is 1:4, or 1 in 4. With this gradient, for every 4 metres you walk horizontally, you will gain or lose 1 metre in height. Or, for every 4 miles you walk horizontally, you will gain or lose 1 mile in elevation. Because this ratio is dimensionless, you may put in whatever units you chose to work with; metres, miles, feet—it doesn't matter as long as you are consistent within the ratio.

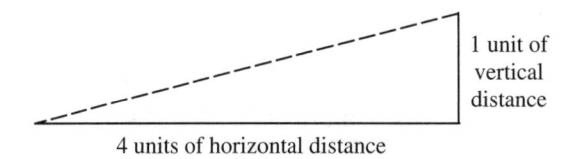

Gradient as a percentage

You might also choose to express the gradient as a percentage. In this case:

$$Gradient = \frac{\text{Change in Elevation }(\Delta E)}{\text{Change in Horizontal Distance }(\Delta D)} = \frac{200m}{800m}(100\%) = 25\%$$

The gradient expressed as a percentage must be interpreted with caution. A 100 per cent gradient means that the elevation change is matched the horizontal distance change, in other words, the land would have a slope of 45°.

Gradient as a fraction

This expression is similar to the gradient as a percentage. Here it is expressed as a fraction.

$$Gradient = \frac{\text{Change in Elevation }(\Delta E)}{\text{Change in Horizontal Distance }(\Delta D)} = \frac{200m}{800m} = 0.25$$

Gradient as a degree

The gradient may also be expressed in degrees of arc. The tangent of an angle refers to the length of the opposite side of a triangle over the length of the adjacent side.

$$Tangent\,\theta = \frac{200m}{800m} = 0.25$$

Tangent θ = 0.25. Using our calculators to find the arctangent of the value will give us the angle. Arctan 0.25 = 14.036°. If expressed in degrees and minutes of arc this would be approximately 14° 02′.

Some Gradient Questions for You

1. An elevation change of 50 m over a horizontal distance of 2500 m would give what gradient (as a ratio, percentage, fraction and degree)?

    Ratio = 1:500
    Percentage = 2%
    Fraction = 0.02
    Degree = 1.1°